PLATING
DESSERT

이은지 LEE EUNJI

파티시에의 꿈을 이루기 위해 열여덟 나이에 프랑스 유학길에 올랐다. 프랑스 국립제과제빵학교INBP에서 블랑제리 과정을, 에꼴 페랑디Ecole Ferrandi에서 파티스리 과정을 수료한 후 프랑스와 뉴욕의 페이스트리 숍과 레스토랑에서 다양한 경험을 쌓았다. 프랑스 인기 TV 프로그램 <Qui Sera Le Prochain Grand Patissier> 시즌 4에서 비유럽인 최초로 준우승을 하며 페이스트리 셰프로서의 입지를 다졌다. 이후 정식Jungsik(뉴욕)에서 선보였던 '베이비 바나나'로 매거진 <Starchefs>에서 'Rising star'로 선정, Art of Presentation 수상, SNS상에서 전 세계인의 관심을 받으며 '이은지'라는 이름을 알렸다. 현재 뉴욕에서 페이스트리 부티크 '리제 LYSÉE'의 오너 셰프로 활동하며 바쁜 나날을 보내고 있다.

@eunji.leeee
@lysee.nyc

2024
Outstanding Pastry Chef and Baker, nominated
Semifinalist by James Beard Awards

2023
Pastry Talents of the Year prize by La Liste 1000
Best New Chefs America prize by Food & Wine
Best Original Kouign Amann winner by NYC
Kouign-Amann Contest

2022
LYSÉE (New York) 오너 셰프

2019
Valrhona C3 대회 North America 준우승
매거진 <Starchefs> 선정 'Rising Star',
'Art of Presentation' 수상

2017
프랑스 TV 프로그램 <Qui Sera Le Prochain Grand
Patissier> 시즌 4 준우승

2016
Jungsik (New York) 디저트 파트 총괄 셰프

2012
Hotel Le Meurice (Paris) 디저트 파트 책임자

2009
Ze Kitchen Galerie (Paris) 디저트 파트 담당

2008
Ecole Ferrandi (Paris) Patisserie 과정 수료,
CAP Patissier 취득

2006
INBP (Rouen) boulangerie 과정 수료,
CAP Boulanger 취득

PLATING DESSERT
플레이팅 디저트

초판 1쇄 발행	2022년 8월 22일
초판 3쇄 발행	2024년 12월 5일

지은이	이은지	주소	경기도 부천시 조마루로385번길 122 삼보테크노타워 2002호
펴낸이	한준희	홈페이지	www.icoxpublish.com
발행처	(주)아이콕스	쇼핑몰	www.baek2.kr (백두도서쇼핑몰)
		인스타그램	@thetable_book
기획·편집	박윤선	이메일	thetable_book@naver.com
디자인	김보라	전화	032) 674-5685
사진	박성영	팩스	032) 676-5685
스타일링	이화영	등록	2015년 7월 9일 제 386-251002015000034호
영업·마케팅	김남권, 조용훈, 문성빈	ISBN	979-11-6426-218-2 (13590)
경영지원	김효선, 이정민		

PLATING
DESSERT

플레이팅 디저트

이은지

LEE EUNJI

더 테이블
THE TABLE

PROLOGUE

이 책은 제가 고민하며 만들어왔던 '플레이팅 디저트'에 대한 레시피와 테크닉에 관한 내용은 물론 학창시절부터 페이스트리 셰프를 꿈꾸면서 뉴욕에 저의 개인 매장 '리제LYSÉE'를 오픈한 지금까지, 꿈을 향해 걸어왔던 길에 대한 소중한 기록이 담겨 있습니다.

레시피만 담긴 책들과는 다르게 이런 개인적인 이야기들을 넣고 싶었던 이유 중 하나는 페이스트리 셰프를 꿈꾸는 학생들, 셰프로 진로를 바꾸고자 하는 분들에게 그동안 개인적으로 메시지를 많이 받아왔기 때문에 이 책을 통해 이런 분들에게 조금이나마 도움을 드리고 싶었기 때문입니다.

만 18세에 프랑스로 떠나 그곳에서 10년간 기술을 다지고 경험을 쌓고, 2016년 뉴욕에서 새로운 삶을 시작해 2022년 뉴욕에서 현재의 개인 매장을 오픈하기까지, 그리고 저의 레시피와 이야기가 담긴 의미 있는 첫 책을 출간하기까지. 디저트 세계에 발을 내디딘 후 벌써 16년이라는 시간이 흘렀으니 참 기쁘고 감사한 일이 아닌가 싶습니다

그간 정말 다양한 에피소드들이 있었고 참 많은 일들이 있었습니다. 제가 여기까지 잘 올 수 있었던 이유는 파티시에로서 제 직업에 대한 열정과 사랑, 제 꿈에 대한 의지, 제 곁에 있었던 좋은 사람들, 사랑하는 가족과 친구들의 응원 덕분이 아니었을까 싶습니다. 의미 있는 첫 책이라 이 분들께 조금이나마 이 책을 통해 감사의 마음을 전하고자 합니다.

첫 시작부터 언제나 든든한 제 1호 서포터즈가 되어준 한국에 계신 부모님과 동생, 그리고 변함없이 늘 저를 응원해주는 한국의 친구들, 프랑스에 있는 가족과 친구들, 모두 감사하고 사랑합니다.

즈 키친 갤러리Ze Kitchen Galerie의 윌리엄 르뙤이William Ledeuil 셰프를 만나지 못했더라면 어떻게 되었을까 싶습니다. 프랑스에 남을 수 있게 기회를 주시고 많은 지원과 가르침으로 저를 이끌어주신 윌리엄 셰프에게 정말 감사드립니다. 르 뫼우리스Le Meurice에 있는 동안 다양한 경험으로 저를 페이스트리 셰프로 더욱 성장할 수 있게 이끌어준 세드릭 그롤레Cedric Grolet 셰프, 저의 베스트 프렌드이자 존경하는 셰프인 막심 프레드릭Maxime Frederic에게도 감사의 인사를 전합니다. 그리고 뉴욕이라는 곳에 올 수 있게 기회를 주신 정식당 임정식 사장님, 존경하고 감사드리며 5년 동안 저를 늘 응원하고 지지해주었던 정식당 팀원들, 가족같은 뉴욕의 사랑하는 친구들에게도 고마움을 전합니다.

2021년부터 약 1년간 '리제' 매장의 오픈을 준비했습니다. 제 이름을 건 뉴욕 매장의 오픈에 도움을 주셨던 모든 분들, 특히 저 하나만 믿고 프로젝트의 첫 시작부터 함께 고생하고 달려와주신 Hand hospitality 이기현 대표님과 직원 분들께 감사드리며, 오픈 전후로 변함없이 늘 열심히 해주고 있는 사랑하는 리제 Pastry팀 Annie, Ashley, Olivia에게도 감사의 인사를 전하고 싶습니다.

마지막으로 저 하나만 보고 프랑스에서 뉴욕으로 날아와 리제의 시작을 함께 하며 고생 많이 하고 있는 남편 마튜 로브리Matthieu Lobry에게 세상에서 가장 사랑하고 고맙다고 말하고 싶습니다. Je t'aime, mon Lobry!

한국의 식재료, 혹은 제철 재료를 사용하고 다양한 조리 방법으로 만들어지는 저의 디저트는 코리안 프렌치 뉴요커라는 저의 정체성과도 많이 닮아있습니다. 세 가지의 문화가 조화롭게 섞인 디저트들을 흥미롭게 봐주셨으면 좋겠습니다. 그리고 이 책이 파티시에를 꿈꾸는 모든 분들에게 꿈과 희망과 응원의 메시지가 되기를 바랍니다.

2022년 7월, 저자 **이은지**

This book is not only about the recipes and techniques for 'Plating Dessert' that I have been contemplating while making but also contains valuable records about the path I have walked towards my dream since I was in school, from dreaming of becoming a pastry chef to opening my own shop LYSÉE, in New York.

Unlike the books that only have recipes, one of the reasons I wanted to include my personal stories was because I personally received a lot of messages from students who dream of becoming pastry chefs and those who want to change their careers as chefs. So I am hoping this book can offer some help for them.

I left for France at the age of 18, where I honed my skills and gained experience for 10 years, and went through a lot of ups and downs, from starting a new life in New York to opening my current shop and publishing this book with my recipes and stories. It's been 16 years since I set foot in the dessert industry, and I am thrilled and thankful.

There have been so many different episodes, and so many things took place. But I believe the reason I was able to come this far was because of my passion and love for my job as a patissier, my will for my dream, the good people who were by my side, and the support of my family and friends. As this is the first meaningful book, I would like to express my gratitude and respect to these people through this book.

I want to thank and express love to my parents and my brother in Korea, who always are my first and most reliable supporters, my Korean friends who root for me unwaveringly, and my family and friends in France.

I wonder what would have happened if I hadn't met Chef William Ledeuil of 'Ze Kitchen Galerie.' I am very grateful to Chef William for giving me the opportunity to remain in France and for guiding me with so much support and teaching. I would also like to thank Chef Cedric Grolet, who has helped me develop further as a pastry chef with his varied experiences while at 'Le Meurice,' and also my best friend and respected chef, Maxime Frederic. Also, I would like to express my gratitude and respect to Chef Jung-Sik Lim, the owner of Jungsik, who allowed me the opportunity to come to New York. I would also like to thank the members of the Jungsik team who have always encouraged and supported me for five years, and my beloved friends in New York who are like family to me.

I have prepared to open the shop, 'LYSÉE' for about one year since 2021. I want to thank everyone who helped to open my New York shop, especially Ki-Hyun Lee, the CEO of Hand Hospitality and his team, who trusted me and went through so much trouble together. I also want to thank our pastry team, Annie, Ashley, and Olivia, for their invariable hard work before and after the opening.

Lastly, I would like to express my love and gratitude to my husband, Matthieu Lobry, who flew to New York from France only having faith in me, started LYSÉE with me, and for going through all the troubles. Je t'aime, mon Lobry!

My desserts, which are made using Korean or seasonal ingredients and using various cooking methods, are very similar to my identity as a Korean French New Yorker. I hope you enjoy the desserts that harmoniously blend the three cultures. And I hope that this book will be a message of dreams, hope, and encouragement to all who dream of becoming a pastry chef.

July 2022, Author **Eunji Lee**

CONTENTS

5

Baby Banana
베이비 바나나

100

초콜릿 헤이즐넛 크럼블	Chocolate Hazelnut Crumble
커피 아이스크림	Coffee Ice Cream
바나나 무스 케이크	Banana Mousse Cake

6

Lotus Flower
연꽃

120

사블레 브르통	Sablé Breton
칼라만시 크림	Kalamansi Cream
만다린 콩포트	Mandarin Compote
세그먼트한 과일	Segmented Fruits
오렌지 & 레몬 & 금귤 콩피	Orange & Lemon & Kumquat Confi
금귤 절임	Steeped Kumquat
핑거라임	Finger Lime
블러드오렌지 소르베	Blood Orange Sorbet
연꽃 머랭	Lotus Flower Meringue

7

New York-Seoul

뉴욕-서울

138

8

Giwa

기와

158

INTRODUCTION
Patissier, LEE EUNJI

열다섯, 파티시에를 꿈꾸다

저는 미술을 좋아하시는 부모님 덕분에(부모님은 대학교 미술 동아리에서 만나 결혼 하셨어요.) 어렸을 때부터 자연스럽게 그림을 그리거나 종이를 접어 무언가를 만드는 예술 활동에 익숙했어요. 어릴 적 사진을 보면 무언가를 그리거나 만들고 있는 사진 이 대부분일 정도였지요. 그러다 15살이 되던 해에 특별활동으로 제과제빵부에 들 어가게 되었는데, 내 손으로 만든 케이크와 빵을 가족과 친구들이 맛있게 먹는 모습 을 보는 게 너무나도 행복했어요. 그래서 집에서도 틈만 나면 과자와 빵을 만들어 친 구의 생일이나 특별한 날에 선물해주었어요. 그러다 우연히 인터넷으로 파티시에의 일상을 다룬 다큐멘터리를 접하게 되었는데, 화려한 디저트를 만드는 파티시에와 그 디저트로 즐거움을 느끼는 손님들의 모습을 보고 파티시에라는 직업에 매력을 느끼게 되었어요. 그때만 해도 파티시에라는 단어조차 생소할 때라 사람들이 '너는 꿈이 뭐니?'라고 물었을 때 '파티시에요.'라고 답하면 돌아오는 대답 대부분이 '그게 뭔데?'였어요. 지금처럼 각광받는 직업도 아니었기에 부모님도 처음에는 반대를 하 셨어요. 하지만 부모님은 제가 쓴 10년 계획표를 보고 제 꿈과 의지가 너무나도 확 고하다는 걸 깨닫고 제 꿈을 응원해주기 시작하셨어요. 물론 지금은 저의 가장 든든 한 1호 서포터이시죠.

유학을 결심하고 처음에는 일본으로 갈 생각이었어요. 제가 일본어를 곧잘 하 기도 했고 한국과 가까운 나라이기도 해서였죠. 하지만 '이왕 갈 거 디저트의 나라인 프랑스로 가보자'라는 생각으로 미식의 나라 프랑스로 떠나기로 결심했어요. 결정 을 한 이후로 불어 학원을 다니면서 언어도 익히고 제과제빵 관련 프랑스 서적을 사 서 전문 용어도 익히며 정말 열심히 공부했어요. 그렇게 시간이 흐르고 2006년 8월, 드디어 프랑스행 비행기에 몸을 싣게 되었어요. 부모님 눈에는 아직도 어리기만 한 딸인데, 저 먼 지구 반대편 나라로 자식을 떠나보내야만 했던 부모님의 눈에 맺힌 눈 물이 아직도 생생해요. 한편 걱정 가득한 부모님과는 반대로 저는 너무나도 행복한 미소를 지으며 '안녕! 나 갈게!'라고 환하게 인사하고 비행기에 올랐어요. 지금 생각 하면 부모님에게는 참 슬펐던 순간이었을 것 같은데 저에게는 그 순간이 제 인생에 있어 가장 행복했고, 설렜던 순간이었어요. 꿈을 향한 첫 걸음이었으니까요. 그렇게 원하는 꿈을 향해 첫걸음을 내딛었어요.

15 Years Old, Dreaming of Becoming a Patissier

Thanks to my parents, who love art (they met at an art club in university and got married), I was naturally accustomed to art activities such as drawing or folding papers to make something since I was young. Even in my childhood photos, most of them are pictures of me drawing or making something. Then when I turned 15, I joined the pastry and baking club as a special activity, and I was so happy to see my family and friends enjoying the cakes and bread I made myself. So, whenever I had time, even at home, I made cookies and bread and gave them to friends on their birthdays or special occasions. Then I happened to encounter a documentary about the daily life of a patissier on the internet, and I was drawn to the profession of the patissier after seeing them making glamorous desserts and the guests enjoying them. At that time, even the word patissier was uncommon; when people asked, 'what is your dream?' and if you answered, 'I want to be a patissier,' most of them would ask back, 'what's that?' My parents were against it at first because it wasn't a popular job like it is now, but I didn't give up on my dream despite their constant opposition. However, when my parent saw my 10-year plan, they realized that my will and dream was strong, so they started to support my dream. But, of course, they are now my most reliable supporters.

When I decided to study abroad, my first consideration was Japan because I spoke Japanese, and it was close to Korea. However, I thought, 'If I'm going somewhere at all, let's go to France, the country of desserts,' and decided to head for France. After making the decision, I studied very hard; I attended French language school to learn the language and bought French pastry and baking books to familiarize myself with the technical terms. And time went by, and in August 2006, I finally got on a plane heading to France. Because I was still just their baby girl to my parents, I still vividly remember the tear in their eyes as they sent me away to a far-off country on the other side of the globe. But, contrary to my worried parents, I bid farewell, happily smiling, saying 'Goodbye, I'm off!!' and hopped on the plane. Thinking about it now, it must have been a very sad moment for my parents, but for me, it was the happiest and most thrilling moment in my life. It was my first step toward my dream. Just like that, I took the first step toward the future I wanted.

제과를 배우기 전에 먼저 제빵을 배우는 것이 좋겠다고 판단해 국립제과제빵학교(INBP, Institut National De La Boulangerie Ptisserie)에서 제빵 과정을 가장 처음 배우게 되었는데 하루하루가 너무 재미있고 즐거웠어요. 이론 시간에는 수업 내용을 완벽하게 이해하기 위해 예습과 복습을 철저히 했고, MP3로 녹음하며 반복해 들으면서 수업 때 놓쳤던 부분을 다시 익히고, 모르는 것이 생기면 적어두었다가 다음날 선생님께 물어보았어요. 지금 생각해보면 힘들었을 법도 한데 그땐 마냥 재미있기만 했어요. 제 눈에 너무 멋진 선생님이 매번 친절하게 가르쳐주셔서 즐겁게 공부했고, 선생님하고 이야기하고 싶어 일부러 질문을 만들어 가기도 했고요. 우리 반은 일본인 친구 한 명을 빼고는 전부 프랑스인이었고, 저는 그중에서 나이도 어린 편이어서 반 친구들이 잘 챙겨주었어요. 주말이 되면 친구들의 초대로 샹파뉴Champagne, 도빌Deauville, 트루빌Trouville, 릴Lille 등 친구들이 사는 지역을 구경하고 그들의 집에서 가족들과 함께 시간을 보내며 놀다 오곤 했어요. 덕분에 저는 프랑스 언어와 문화를 빨리 습득할 수 있었어요. 이렇게 좋은 사람들을 만날 수 있어 참 감사했죠.

졸업이 다가올 때쯤 인턴십 프로그램을 알아보고 있었는데 친구 한 명이 스트라스부르Strasbourg라는 도시를 아주 멋진 곳이라고 추천해 가보니 정말 멋진 도시였어요. ('Au Pian de mon Grand-Père'라는 유명한 빵집도 가보았는데 제가 여태껏 먹었던 바게트 중 단연 최고였어요.) 계속 일을 하기 위해서는 학생 비자를 연장해야 해서 학교를 다녀야 했는데 이왕 다녀야 하는 거 불어 자격증에 도전해보자는 마음이 들어 근처 대학교의 어학원을 등록했어요. 빵집들은 보통 아침 일찍 빵을 구워야 하기 때문에 밤이나 새벽에 출근을 하는데 저 또한 예외 없이 밤 12시부터 오전 8시까지 일하고 세상 가장 맛있다는 갓 구운 빵을 입에 물고 학원으로 가 오후까지 수업을 들었어요. 그리고 나서 그날의 복습과 숙제를 하고 이른 저녁에 잠들어 다시 밤에 일어나 출근을 했죠. 제 또래들과는 조금 다른 시간에 살았지만 하루하루 너무 재미있고 신이 났어요. 당시 같이 수업을 들었던 언니, 오빠들이 저에게서 항상 고소하고 달콤한 냄새가 난다고 좋아해주던 기억이 나요.

어느 날 저에게 할아버지뻘이었던 가게 오너 셰프님이 어시스트를 제안하셨어요. 평소에도 저를 예뻐해주셨었는데 덕분에 저는 생활비도 조금 더 벌 수 있게 되었고 빵 반죽도 더 많이 만져보고 배울 수 있게 되었죠. 그렇게 얼마간의 시간이 흐르

I decided that it would be better to learn baking before learning pastry, so I first started with the baking course at INBP (Institut National De La Boulangerie Patisserie) and each and every day was so much fun and enjoyable. For the theory class, I thoroughly prepared and reviewed to understand the lecture perfectly. I recorded the lecture and listened over and over to absorb the parts I missed in class, and if I didn't understand something, I wrote it down and asked the next day. When I think about it now, it must have been difficult, but it was just all fun back then. The professor, who was so handsome in my eyes, taught so kindly all the time that studying was fun, and I made up questions on purpose, wanting to talk with the professor. All my classmates were French except for one Japanese friend, and my classmates took good care of me because I was the youngest among them. On weekends, they invited me to visit the towns where they lived, such as Champagne, Deauville, Trouville, Lille, etc. and spent time with their family at their homes. Thanks to them, I could learn the French language and culture quickly. I am so grateful to have met such cordial people.

As graduation approached, as I was looking for an internship, one of my friends recommended Strasbourg, saying it was a fantastic place to go to; it turned out to be a truly wonderful city. (I went to a famous bakery called 'Au Pian de mon Grand-Père' and they had by far the best baguette I have ever eaten.) In order to continue working, I had to attend a school to extend my student visa. And I thought if I was going to go to a school, I might as well get the French language certificate, so I enrolled in a language school at a nearby university. Bakers usually start working at night or dawn because they have to bake early in the morning; so I, too, worked from 12 a.m. to 8 a.m. without exception and went to school with freshly baked bread, which what they say is the best-thing-in-the-world, in my mouth and attend classes until the afternoon. Then after reviewing the day's lecture and doing my homework, I went to sleep early in the evening, woke up at night, and headed out to work. I lived in a slightly different time zone than my peers, but every day was so fun and exciting. I remember my classmates saying that they loved how I always smelled like fresh-baked bread.

One day, the owner-chef of the bakery, who was as old as my grandfather, suggested becoming an assistant. He always cared for me, and thanks to him,

고 저의 최종 목표인 파티시에를 위해 이 멋진 도시와 작별인사를 하고 파리로 떠나게 되었어요.

2008년 파리 에꼴 페랑디Ecole Ferrandi에 입학해 배운 모든 것들이 너무나도 즐겁고 신기하고 소중했어요. 하지만 제과 과정을 수료하고 작은 페이스트리 숍에서의 짧은 인턴 생활이 끝난 후 어려운 시기가 찾아왔어요. 페이스트리 숍에 정식으로 고용되어 일하고 싶어 워킹 비자를 해줄 수 있는 곳을 찾아야 했는데 외국인을 고용하려는 곳을 찾는 게 생각보다 너무 어려웠죠. 파리부터 저 멀리 남부 지역까지 100장이 넘는 이력서를 돌리고, 편지를 쓰고, 직접 찾아가기까지 했지만 소용없었어요. 그렇게 비자 만료일은 점점 다가오고 허탈함과 불안함이 극에 달했을 때 정말 한 줄기 빛처럼 '즈 키친 갤러리Ze Kitchen Galerie'라는 파리의 1스타 미슐랭 레스토랑을 알게 되었어요. 다행히 이곳은 외국인 워킹 비자에 대해 오픈된 마인드를 가지고 있었고 제가 일을 잘하기만 하면 바로 해줄 수도 있다고 했어요. 저에게는 더없이 소중한 기회였기에 최선을 다해 정말 열심히 일했고, 다행히 비자가 만료되기 전에 워킹 비자를 받을 수 있게 되었어요. 저에게는 첫 레스토랑 경험이었는데 디저트 파트에 담당자가 없어 인수인계도 받지 못하고 저 혼자서 모든 것을 처리해야 했어요. 하루에 120~130인분의 디저트를 만들어내야 했기 때문에 혼자 고군분투했지만, 그래도 오너 셰프 윌리엄 르되이William Ledeuil처럼 좋은 사람들과 일하면서 많은 것을 배우고 경험할 수 있었어요.

윌리엄 셰프는 항상 신선하고 질 좋은 식재료를 경험할 수 있게 해주었어요. 덕분에 몰랐던 식재료에 대해 알게 되었고 스스로 이것저것 테스트도 해보고, 어깨너머로 배운 요리에 사용되는 조리 방식을 디저트에 접목시켜보기도 했어요. 셰프가 작업하던 쿡북에서 디저트 파트로 참여하기도 했고요. 이렇게 이곳에서 3년을 일하면서 성장할 수 있는 좋은 경험들을 많이 한 것 같아요. 하지만 저는 정통 페이스트리에 대한 열망이 있었기에 전문 페이스트리 셰프들 사이에서 파티스리에 관한 전문적인 조언과 가르침을 받고 더 성장하길 원했어요. 윌리엄 셰프는 제가 좀 더 오래 남아 있기를 원했지만 저의 결정을 이해하고 응원해주셨어요. 좌절했던 순간 저에게 손을 내밀어준 윌리엄 셰프에 대한 고마움은 항상 마음속에 있어요. 덕분에 프랑스에 더 머물 수 있었고 페이스트리 셰프로서 좋은 경험과 실력을 쌓아가며 성장할 수 있었고 사랑하는 남편도 만나게 되었으니까요.

즈 키친 갤러리 오너 셰프 윌리엄 르되이와 함께
With William Ledeuil,
the owner-chef of 'Ze Kitchen Galerie'

I was able to earn a little more, and it was also a good opportunity for me to get my hands on the bread dough and learn more. After some time, I bid farewell to this wonderful city and moved to Paris for my ultimate goal- a patissier.

In 2008, I entered École Ferrandi in Paris, and everything I learned was fun, engaging, and precious. After completing CAP Patisserie Program and a short internship at a small pastry shop, I had to face a difficult time. I wanted to be hired full-time at a shop, so I had to find a shop that could provide a work visa, but finding a place where they would employ foreigners was harder than I thought. From Paris to the far South, I sent over 100 resumes, wrote letters, and even visited them, but nothing worked. As the expiration date of my visa was approaching and my feelings of disappointment and anxiety peaked, I discovered a 1 Star Michelin restaurant in Paris called 'Ze Kitchen Galerie' like a ray of light. Fortunately, this place had an open mind about work visas for foreigners, and they said they could process it right away if I did a good job. It was an invaluable opportunity, so I worked really hard on, and fortunately, I was able to get the work visa before my visa expired. I had to handle everything by myself, even though it was my first restaurant experience because there was no person in charge of the dessert section to hand over the duties. I struggled alone because I had to make 120 to 130 servings of desserts a day, but I was able to learn and experience many things while working with good people like the owner-chef William Ledeuil.

Chef William always allowed me to experience fresh and high-quality ingredients. Thanks to him, I learned about ingredients I wasn't aware of, tested numerous things on my own, and incorporated the cooking methods used in a cuisine I learned over the shoulder into my desserts. I even participated in the dessert part of the cookbook he was working on. I believe I had a lot of good experiences that allowed me to grow while working here for three years. However, because I had a passion for traditional pastries, I wanted to develop further by getting professional advice and teaching on the pastry from expert pastry chefs. Chef William wanted me to stay longer, but he understood my decision and supported me. I am always grateful to Chef William for reaching out when I was frustrated. Thanks to him, I was able to stay in France longer and grow as a pastry chef with fabulous experiences and skills and met my beloved husband.

즈 키친 갤러리의 주방
The kitchen of 'Ze Kitchen Galerie'

수셰프 막심 프레드릭과 함께
크로캉 부슈, 초대형 생토노레를 작업하며

With Sous Chef Maxim Frederick,
working on croquembouche and
a supersized Saint-Honoré

2012년 4월, 르 메우리스Le Meurice 호텔에서 일하게 되었어요. 당시 파리에는 palace급(5성급 이상) 호텔이 5개 정도 있었는데 르 메우리스도 그중 하나였고, 입사하기도 어려웠지만 그 안에서 살아남기는 더 어려운 악명 높은 곳이었죠. 하지만 저는 넓고 쾌적한 페이스트리 주방, 그리고 새하얀 셰프복을 입고 각자의 작업에 몰두하고 있는 수십 명의 셰프들의 모습을 보며 놀이공원에 온 아이처럼 설렜어요.

처음에는 카미유 레스크Camille Lesecq라는 전설적인 페이스트리 셰프가 계셨지만 아쉽게도 제가 일한 지 반년도 지나지 않아 그만두셨고, 당시 수셰프였던(지금은 제과 일을 하는 사람이라면 누구나 다 아는 세계적으로 유명한 셰프) 세드릭 그롤레Cedric Grolet가 페이스트리 파트 총괄 셰프로 올라가게 되었어요. 페이스트리 팀은 당시 인턴 포함 스무 명 정도였고, 한 팀은 당시 3스타였던 '파인 다이닝' 팀, 그리고 다른 한 팀은 티타임과 뱅큇, 룸서비스, 일반 레스토랑 등 파인 다이닝 외에 모든 것을 맡는 '달리' 팀 이렇게 두 팀으로 나뉘어져 있었어요. (그 일반 레스토랑의 이름이 예전에 살바도르 달리가 자주 찾았다고 해서 붙여진 이름인 'Dali'입니다.)

저는 이미 레스토랑 파트를 경험했기 때문에 플레이팅 디저트보다 좀 더 부티크 제품에 가까운 달리 팀에서 쭉 일했어요. 덕분에 정말 다양한 작업을 할 수 있었고, 또 많이 성장할 수 있었어요. 한편으로는 너무나도 치열했고, 엄격했고, 일하는 시간도 어마어마했지만 이곳에서 일하는 매일이 행복하고 즐거웠죠. 사원급 셰프(commis)로 시작해 1년 뒤에는 선임 사원급 셰프(demi de partie)로, 그 다음 해에는 주임급 셰프(chef de partie)로 승진했어요. 방대한 업무로 원래는 두 명의 주임급 셰프로 이루어진 팀은 저 혼자서 맡게 되었고, 당시 수셰프였던 막심 프레드릭Maxime Frederic과 함께 달리 팀을 이끌면서 경영에 대해서도 많이 배울 수 있는 계기가 되었어요. 대부분의 셰프가 프랑스인이었기 때문에 부담과 스트레스가 많았는데 그때마다 제 자신보다 저를 더 믿어주었던 세드릭 셰프와 막심 셰프 덕분에 스스로 더 단단해질 수 있었죠. 르 메우리스에서의 시간은 저에게는 페이스트리 셰프로서 갖추어야 할 기술뿐만 아니라 경영, 리더십 등 다방면으로 성장할 수 있었던 소중한 경험이었어요.

In April 2012, I got to work at the Le Meurice Hotel. Back then, there were about five palace-class (5 Star hotels) in Paris; one of them was Le Meurice, which was known to be notoriously difficult to get in, but more demanding to survive. However, I was thrilled like a child at an amusement park to see such a spacious and pleasant pastry kitchen with dozens of chefs wearing chef jackets white as snow, each immersed in their work.

At first, there was a legendary pastry chef named Camille Lesecq, who unfortunately left half a year after I started. And Cedric Grolet, who was then a sous chef (now a world-famous chef to anyone in the pastry industry), became the head chef of the pastry part. The pastry team consisted of about 20 people, including interns at the time, and they were divided into two teams: One team belonged to the 'fine dining' team, which back then had 3 Star, and the other team at 'Dali,' who took care of everything except the fine dining, such as tea time, banquet, room service, and general restaurants. (The name of the general restaurant is 'Dali,' a name given due to Salvador Dali's frequent visits in the past.)

I have already experienced the restaurant part, so I had worked with team Dali for some time, which is more of boutique products than plated desserts. Thankfully, I was able to try a lot of different work, and I was able to grow a lot. On the other hand, it was incredibly intense, strict, and had tremendous working hours, but every day I worked here was happy and enjoyable. At first, I started as a commis, and after a year, I was promoted to demi chef de partie and the following year to chef de partie. Team Dali was initially led by two chefs de partie, but as I got promoted, I had to lead the team single-handedly. And due to its extensive work, it was an opportunity to learn a lot about management while leading team Dali with Maxime Frederic, the sous chef at the time. I was under a lot of pressure and stress because most of the chefs were French, but I was able to strengthen myself thanks to Chef Cedric and Chef Maxime, who believed in me more than I did myself. The time at Le Meurice was a valuable experience that allowed me to develop not only the skills I should have as a pastry chef but also grow in various fields such as management and leadership.

세드릭 그롤레와 함께
With Cedric Grolet

그렇게 2년이라는 시간이 흐르고 세드릭 셰프가 저에게 호텔 수셰프의 타이틀과 함께 곧 오픈할 르 메우리스 제과점의 총괄 제과장이 되어 달라는 제안을 했어요. 당시 뉴욕 미슐랭 2스타였던 정식(뉴욕)의 페이스트리 셰프 제안도 받았기에 선택의 기로에 섰어요. 너무나도 어려운 선택이었죠. 긴 고민 끝에 저는 10년간의 프랑스 생활을 접고 정식 뉴욕의 총괄 제과장으로 일하게 되었어요. 인센가 세가 좋아하는 뉴욕에서 일해보고 싶다는 생각도 있었고, 정식당은 개인적으로도 좋아하는 곳이라 꼭 가보고 싶었던 곳이기도 했었고, 프랑스에서 배웠던 것들을 바탕으로 한국의 식재료와 현지의 재료를 조화시켜 오직 저만의 스타일로 풀어낸 디저트를 전세계 사람들이 모여 있는 뉴욕에서 테스트해보고 싶은 마음도 있었죠. 그렇게 2016년부터 2021년까지 5년이라는 시간을 정식당의 일원으로 큰 책임감을 가지고 임했습니다.

정식당이 아니었다면 저는 아마 계속 프랑스에 남아있지 않았을까 싶어요. 그때 제안 주시고 좋은 기회 주신 임정식 셰프님께 항상 감사하는 마음을 가지고 있어요. '좋은 기회가 왔을 때 경험해 보자'라는 마음이 들었고, 또한 누군가의 디저트를 만드는 게 아닌 한 레스토랑의 총괄 제과장으로 여태까지 쌓아왔던 경험들과 저의 아이디어를 바탕으로 오롯이 저만의 디저트를 만들어보고 도전해보고 싶다는 마음이 컸던 것 같아요. 만 29세, 한국 나이로 30대의 시작을 새로운 곳. 새로운 환경에서 나만의 디저트를 만들어보자는 도전 정신과 모험심이 가장 큰 이유였던 것이죠. 물론 10년 간 있었던 곳을 떠난다는 게 쉽지 않았고, 뉴욕에 와서 여러 가지 힘든 일이 많기도 했지만 후회 없이 잘한 선택이었다고 생각해요.

Two years have passed, and Chef Cedric offered me to be the executive pastry chef of the Le Meurice Pastry Shop, which was about to open, along with the title of sous chef at the hotel. I was also offered a pastry chef at Jungsik, a 2 Star Michelin restaurant in New York at the time, which had me at a crossroads of choice. It was such a difficult choice. After much deliberation, I wrapped up my ten years of living in France and started as an executive pastry chef at Jungsik in New York. I wanted to work in New York someday, the city I love, and Jungsik is a place I admire, where I yearned to go. I also wanted to test out the desserts created in my own style by harmonizing Korean and local ingredients based on what I learned in France in New York, where people from all over the world gather. So, I worked from 2016 to 2021 with great responsibility as a member of Jungsik for five years.

If it wasn't for Jungsik, I think I probably would have continued to stay in France. I am always grateful to Chef Jung-Sik Lim, who offered me the job and gave me a good opportunity. I thought, 'Let's try it when a good opportunity is here,' and I had a strong will to make my desserts based on my own ideas and heritage, as well as my experience as the head pastry chef of a restaurant rather than making someone else's desserts. The biggest reason was the spirit of challenge and adventurousness to make my dessert in a new place, new environment, at 29 years old, soon to begin my 30s. Of course, it was not easy to leave the place I had been in for 10 years, and there were many difficult moments in New York, but I can say it was a good choice without regrets.

'전사'라는
별명을 얻다

프랑스에서는 일 잘하는 사람을 'machine de guerre'라고 표현해요. 직역하면 '전쟁 기계', 의역하면 '전사'라는 뜻인데 제가 손도 빠르고 배우는 속도도 빨라 주어진 시간에 많은 일을 정확하게 해서 붙여진 별명인 것 같아요. 그때는 많은 일과 장시간 업무를 감당할 수 있을 만큼 체력도 좋았고, 제가 할 일이 무엇인지, 그리고 어떤 기준으로 만들어야 하는지도 정확히 픽익했어요. 열성과 자무심을 가지고 최선을 다해 정말 열심히 일했어요. 셰프의 신임을 얻어 주말에는 저 홀로 책임자로 일하기도 했고요. 저의 성장 배경에는 저를 믿고 이끌어준 셰프와 수셰프, 그리고 저를 잘 따라준 팀원들이 있었어요. 자리가 사람을 만든다고 자연스럽게 리더십도 생기게 되었고 그 결과 한 단계 한 단계 승진도 하게 된 것 같아요.

프랑스 TV
서바이벌 프로그램에서
비유럽인 최초로
준우승을 하다

어느 날 프랑스 인기 서바이벌 프로그램인 〈Qui Sera Le Prochain Grand Patissier(다음엔 누가 위대한 파티시에가 될 것인가)〉 시즌 4에 지원해보겠냐는 연락을 받았어요. 프랑스에 있을 때 챙겨 보았던 프로그램이기도 했고, 뉴욕에서 힘들었을 때, 마음을 다잡고 싶을 때마다 다시 보기를 했던 TV 프로그램이었어요.

처음에는 뉴욕에서 일하고 있어 참여하기 힘들 것 같다고 거절을 했는데 거절하고 나서도 자꾸 마음속에 아쉬움이 남더라고요. 제작진은 체류증이 있으면 뉴욕에 있어도 참가하는 데 문제가 없다고 저를 계속 설득했어요. 망설이며 정식당과 팀원들에게 이야기를 꺼냈는데 이런 좋은 기회가 어디있냐며 나가보라고 용기를 북돋아주었어요. 당시 저는 첫 총괄 제과장으로 일하고 있던 시기라 '내가 지금 잘 하고 있는 걸까?'라는 생각이 자주 들었고, 나 자신을 한번 테스트해보고 싶은 생각이 들어 참가 지원서를 쓰고 화상 채팅으로 면접을 보았어요.

그 후 파리에서 열린 두 가지 주제의 디저트 테스트까지 최종 통과하여 top 9 최종 멤버에 들게 되었어요. (이때 테스트를 보러 근무 휴일인 이틀 동안 파리에 정말 짧게 다녀왔던 기억이 나요.) 제가 너무 좋아하던 프로그램에 참여하게 되다니 꿈만 같았죠. 비

Acquired the Nickname 'Machine de Guerre (Warrior)'

In France, a person who works hard and does a good job is described as a 'machine de guerre.' It literally translates to a 'war machine' and liberally a 'warrior.' I think I was given the nickname because I work quickly, learn and absorb fast, and I am able to do a lot of work precisely in a given time. At that time, I was physically strong enough to handle a lot of work and long hours, and I knew exactly what I was going to do and what standards I had to make. I worked really hard with passion and pride, doing my best. I gained the chef's trust and worked on the weekends alone as a manager. In the background of my growth, there was the chef and sous chef who believed in me and guided me, and there were team members who followed me well. I believe the responsibility naturally matured my leadership, and as a result, I got promoted one step at a time.

Became The First Non-European to Finish as Runner-up in The French TV Survival Contest

One day, I received a call asking if I would apply for season 4 of <Qui sera le prochain grand patissier (Who Will Be The Next Great Patissier)>, a popular French survival show. It was a show I watched while I was in France, and I used to watch it again whenever I wanted pull myself together when I was having a hard time in New York.

At first, I refused because I was working in New York, so I thought it would be difficult to participate, but even after I declined, I kept feeling wistful. The production team continued to persuade me that if I had a residence permit, there would be no problem in participating even if I was in New York. I hesitated and talked with the restaurant and team members, but they told me what a great opportunity this was and encouraged me to try it out. At the time, I was on my first duty as the executive chef, and I often thought, 'Am I doing well?' I wanted to test myself, so I submitted the application for participation and interviewed through video chat.

After that, I passed the dessert contest of two themes in Paris and was placed in the top 9 finalists. (I remember taking very short trips to Paris for two days on my off days for the preliminaries.) It felt like a dream to be a part

유럽인 최초로 참여하고 준우승까지 차지했고 멋진 호텔에서 수많은 저명한 셰프들, 그리고 멀리 한국에서 온 가족들 앞에서 제 디저트를 선보이고 결승전을 치루기까지 참 믿기지 않을 만큼 꿈같고 감사한 경험이었어요.

재미있는 에피소드로 프랑스 각 지역의 특산 과자를 맞추는 깜짝 테스트가 있었는데, 한 명씩 공간에 들어가서 진열되어 있는 제품들을 보고 지역과 이름을 맞추는 퀴즈였어요. 예상 외로 프랑스 친구들을 제치고 제가 가장 많이 맞춰 그 곳에 있던 사람들이 전부 프랑스인들보다 더 잘 안다며 놀라고 재미있어 했던 기억이 있어요. 또 누아르무티에Noirmoutier라는 섬에 위치해 있는 라 마린La Marine이라는 2스타 미슐랭 레스토랑에서 2박 3일간 머물면서 셰프들과 겨룬 적이 있는데, 키친 전체가 통유리로 되어 있어 밭이 보이는 풍경을 앞에 두고 햇빛을 받으며 그 섬의 특별한 식재료들을 가지고 작업해 볼 수 있어서 너무 신나고 재미있었어요. 그리고 바다 바로 앞의 돌 위에 앉아 노트를 가지고 스케치하며 아이디어를 구상하는 시간이 있었는데 파도가 치는 드넓은 바다를 보니 날씨도 좋고, 너무 좋아하는 굴 생각도 하면서 자연스럽게 누워 엎드린 자세로 구상을 하다 보니, 같이 경쟁하던 친구가 '은지, 자면 안 돼!'라고 말해서 재미있었던 기억이 나요. 2박 3일간 자연 속에 머물면서 앞에는 바다, 뒤에는 밭이 펼쳐져 있는 너무나도 멋진 곳에서 그 지역의 특산물(보통 요리에 쓰는 식재료)을 사용하여 즉흥적으로 디저트를 만들고(저는 감자와 메밀을 응용한 디저트를 만들었어요.) 셰프의 맛있는 음식을 맛보며 많은 이야기도 나누었는데 이런 특별한 모든 경험들이 너무나도 좋았어요.

가장 기억에 남는 메뉴는 트롱프 뢰유trompe l'œil 미션에서 만든 '호롱등Ma Lampe a Bougie'이라는 디저트인데요. 동양적인 느낌을 표현하는 디저트를 만들고 싶어 호롱불을 떠올려 표현했어요. 겉에는 초콜릿으로 만든 호롱을 덮는 케이스, 그리고 케이스를 열면 호롱 모양의 케이크가 나오는 디저트였어요. 다행히 원하는 대로 잘 표현되어 심사위원들 모두가 너무나도 좋아했고, 프레젠테이션 테이블에서 어떤 게 디저트고 어떤 게 물건인지 찾을 수 없다고 놀라워 할 정도로 비주얼도 좋고 초 모양의 케이크도 완벽하다는 극찬을 들었죠. 크리스토퍼 아담Christophe Adam 셰프는 저에게 갑자기 절을 하며 무릎을 꿇을 정도로 너무 맛있다고 극찬을 해주어 정말 감격스러웠어요. 시간이 한참 지나 〈Fou de Patisserie〉라는 매거진에 '셰프가 여태까지 가

of a program I loved so much. It was an incredible dreamlike and grateful experience to become the first non-European to participate and even be the runner-up, to present my desserts and be in the finals in such a beautiful hotel in front of many famous chefs, and my family who came far from Korea.

There was an interesting episode; they gave a surprise quiz where they displayed local snacks from different regions in a designated room, and the participants would go in one at a time and guess the region and name of the snack. However, I answered more of them correctly, far more than the French chefs did, so I remember everyone there being surprised and amazed and saying that I knew more than the French. On one occasion, I competed with the chefs while staying at a 2 Star Michelin restaurant called La Marine on the island of Noirmoutier for three days and two nights. The entire kitchen was made of glass, making it exciting and fun to be able to work with the special ingredients of the island in the sunlight coming in through the window in front of the scenery of the field. And there was a time when we sat on a rock in front of the ocean and sketched out ideas in the notebook. I was looking at the vast wavy ocean, the weather was nice, thinking about the oysters I like very much, and naturally, I was lying on my stomach, thinking about my menu. It was funny because one of the competing chefs saw me and said, 'Eunji, you can't fall asleep!' Staying in nature for three days and two nights, in a gorgeous place with the ocean in the front and fields in the back, using local specialties (usually used in cooking) to improvise desserts (I made a dessert using potatoes and buckwheat) and discuss while tasting the chef's delicious food; all these special experiences were so much fun.

The most memorable menu is a dessert called 'Ma Lampe a Bougie (candle lamp)' made for the Trompe l'oeil mission. I wanted to make a dessert that conveyed an oriental look, so I thought of a lantern and expressed it. It was a dessert with a case made of chocolate that covers the candle, and when you open the case, you'll see a candle-shaped cake. Fortunately, it came out just the way I wanted it to, and all the judges loved it so much. During the presentation, I heard rave reviews about how good it looked and how perfect the candle-shaped cake was, to the point where they were amazed and even

장 맛있게 먹었던 기억에 남는 디저트는 무엇이냐?'는 인터뷰 질문에 제 이야기와 함께 이 디저트 이야기가 실려서 놀라기도 했는데 정말 감사한 일이었죠. 그때 미션 시간이 꽤 길었던 걸로 기억되는데, 그 미션이 저에게는 가장 중요하다고 생각되어 최대한의 집중력으로 온 힘을 다해 시간 안에 디저트를 완성했고, 촬영이 끝나고 쓰러졌던 기억이 있어요. 그 미션에서 공동 우승을 한 친구와 함께 다시 한번 '머랭'을 주제로 번외 미션을 치렀는데, 결국 제가 최종 우승을 하면서 다음 미션에 바로 출전할 수 있는 특권이 생겼죠. 이 미션은 저라는 사람을 '파티시에 이은지'로 각인시키는 계기가 되었어요. 제가 표현하고자 했던 느낌 그대로 디저트가 나와서 만족했고, 셰프들에게 인정도 받았으며, 끝나고 쓰러지기도 했기에 가장 기억에 남는 메뉴예요.

TV 프로그램에 출연했던 경험은 저를 더 성장할 수 있게 만든 계기가 되었어요. 아무나 경험할 수 없는 것이었고 미션을 통해 저 스스로도 더 성장했으며, 좋은 동료들과 멋진 셰프들도 많이 만날 수 있었고 쉽게 갈 수 없었던 키친에서 경험하고 배우고 느낄 수 있었던 꿈같은 시간이었죠. 그리고 큰 자신감도 얻었고요. 사실 저는 꼭 우승을 하겠다는 마음보다는 나의 색이 담긴 디저트들을 보여주었으면 좋겠다는 마음으로 임했거든요. 프랑스 유학 시절부터 우러러 보았던 셰프들이 나의 디저트를 어떻게 받아들일지 궁금했어요. 그래서 저는 미션마다 그냥 제가 하고 싶은 것들을 했어요. 사람들에겐 모험과 도전을 하는 이은지로 받아들여졌고, 이러한 과정들 속에서 셰프들에게 칭찬을 받고 인정을 받았죠. 그리고 '내가 내 길을 잘 가고 있구나. 잘 하고 있구나.'라는 큰 용기와 위로, 자신감을 얻었어요. 방송이 나간 후 정식당에 유럽 손님들은 물론 다양한 국적의 손님들이 많이 찾아와주셨어요. 그리고 각국의 응원 메시지도 정말 많이 받았고요. 이것이 저에게 또 한 번 뉴욕에서 달릴 수 있던 큰 에너지가 되었지요. 정말 너무나도 감사했던, 평생 잊지 못할 기억으로 남을 거예요.

said that they could not tell which was the dessert and which was an object on the table. I was so touched because Chef Christopher Adam suddenly bowed down on his knees and praised how just so delicious it was. After a long while, the magazine 'Fou de Patisserie' published an interview with him, asking him, 'What is the most memorable and delicious dessert you've ever had?' I was surprised and very grateful to find the story of myself and this dessert. I remember that the time given for the mission was quite long. Still, I knew this mission was the most important to me, so I did my absolute best to complete the dessert in time with maximum concentration, and I recall collapsing after filming. With the chef who shared the win in that mission, I once again had an extra mission with the theme of 'meringue.' Ultimately, I won and had the privilege to participate in the next mission. This mission has allowed me to establish myself as the Patissier Eunji Lee. I was satisfied that the dessert came out exactly the way I wanted to express it, the chefs acknowledged me, and I even collapsed after working with all my might; for these various reasons, this menu is the most memorable for me.

The experience of appearing in a TV show allowed me to grow further. It was something not everyone could experience. It was such a grateful and dream-like time where I grew more through the missions, met a lot of good colleagues and incredible chefs, and went into the kitchens I never dreamed I would experience, learn, and perceive. And I gained a lot of confidence. In fact, I participated with the desire to show desserts of my own color rather than the desire to win. I was curious about how the chefs I sincerely looked up to ever since I studied in France would acknowledge my desserts. So I just did whatever I wanted to do for every mission. People recognized me as the Eunji Lee who adventures and challenges, and while doing so, I was praised and recognized by the chefs, which meant a lot to me. And I thought, 'I am walking my path well. I'm doing good.', which let me gain a lot of courage, comfort, and confidence. After the show aired, not only European guests but also people of various nationalities visited the restaurant, Jungsik. And I received so many supportive messages from different countries. This also gave me great energy to keep moving on in New York again. It's an experience and memory I will never forget for the rest of my life.

club des sucres 미션 우승 후
특별 심사위원이었던 셰프들과 찍은 기념 사진
Commemorative photo taken with the chefs
who were the special judges
after winning the Club des Sucres mission

© Dan Ahn

베이비 바나나
Baby Banana

시그니처 메뉴, '베이비 바나나'의 탄생

2016년 뉴욕으로 와 뉴요커들의 입맛을 파악하기 위해 시장 조사를 하러 돌아다닐 때, 한국에서도 유명한 매그놀리아magnolia 베이커리의 바나나 푸딩을 참 맛있게 먹었어요. 개인적으로 바나나우유, 바나나맛 과자 등 바나나를 이용한 스낵들을 무척 좋아하기도 했고, 당시에는 한국에서 초코파이나 몽쉘 등 바나나를 이용한 과자들이 많이 출시되었죠. 저는 마트에 구경하러 가는 걸 좋아하는데 어떤 마트를 가든 과자 코너를 유심히 살펴봐요. 요즘엔 뭐가 나오는지, 뭐가 제일 인기가 많은지 등 다양한 트렌드들을 볼 수 있기 때문이에요. 프랑스에서는 바나나를 디저트에 사용하는 일이 굉장히 드문데, 미국과 아시아는 바나나를 이용한 디저트 문화가 대중적으로 자리 잡고 있어요. 그래서 이 모든 아이디어들을 바탕으로, '뉴요커들이 사랑하는 바나나 푸딩을 나만의 스타일로 한번 만들어보자.('만들어서 내가 매일 먹어야지'라는 생각도 있었습니다.)'라는 게 첫 시작이었죠.

정식당에서는 10가지의 디쉬로 이루어진 시그니처 테이스팅 메뉴 중, 프리 디저트 - 메인 – 포스트 디저트(또는 애프터 디저트) – 프티 프루 이렇게 네 가지의 디저트 코스가 있었어요. 그중 포스트 디저트로는 가공된 퓌레나 첨가물을 쓰지 않고 생바나나가 듬뿍 들어간 바나나 크레뮤, 둘세 가나슈, 바나나 케이크 등으로 이루어진 바나나 푸딩을 선보였어요. 작은 스푼으로 떠먹는 얼마 되지 않는 양이었지만 반응은 폭발적이었어요. 바나나 푸딩만 추가로 시키는 손님도 있었고, 따로 팔지 않냐고

The Birth of Signature Menu, 'Baby Banana'

When I came to New York in 2016 and visited different places to do market research to understand New Yorkers' tastes, I really enjoyed the banana pudding from Magnolia Bakery, also famous in Korea. Personally, I love banana-based snacks such as banana milk and banana-flavored puffs, and at that time, a lot of banana-based snacks such as ChocoPie* or Monshell* cakes were released in Korea. I always like to browse around in the marts, but I particularly look closely at the snack section wherever I go. It's because you can see various trends, such as what's new or most popular, etc. Bananas are rarely used in desserts in France, but desserts using bananas are quite popular in the United States and Asia. So, based on all these ideas, I decided to make my own version of banana pudding, which New Yorkers love (and also thought I would make and eat every day myself), was the starting point.

* Korean chocolate snacks with marshmallow or cream-filled cakes covered in chocolate.

At Jungsik, New York, among the signature tasting menu consisting of 10 dishes, there were four-course dessert menus: pre-dessert - main - post-dessert (or after dessert) - petit fours. Among them, as a post-dessert, a banana pudding made with banana cremuex with plenty of fresh bananas, dulcey ganache, banana cake without processed puree or additives was introduced. It was a small portion eaten with a small spoon, but the reaction was explosive. Some customers would order extra banana pudding only, and some asked if it was sold separately. Thankfully, the guests adored it so much that Chef Jungsik Lim suggested making it the main dessert. So, I agonized over arranging the dessert on the plate to approach customers better with more impact and make it a precious gift in their memories. Then, while I was browsing the fruit section at the market, a baby banana caught my eyes, and I thought, this is it! I thought I would apply the skills I learned and trained in France to make an authentic baby banana replica. Hence, I made a silicone mold in the shape of a banana myself, made a yellow base sauce mixed with cocoa butter and chocolate and dipped in it, and sprayed it with darker color to express the contrast and texture. To make it more realistic, I carefully hand-painted each one using cocoa powder. It looks like one simple menu when plated, but it's an artisan dessert that takes three days to make.

묻는 손님도 있었어요. 감사하게도 손님들이 너무 좋아해주셔서 임정식 사장님께서 한번 메인 디저트로 만들어보는 게 어떻겠냐고 하셨죠. 그래서 어떻게 하면 손님들에게 더욱 임팩트 있게 다가갈 수 있을지, 손님들의 기억 속에 소중한 선물처럼 남을 수 있을지 등 플레이팅을 풀어내는 방법에 대해 많은 고민을 하고 있었어요. 그러던 중, 마켓에서 과일 코너를 구경하고 있는데 베이비 바나나가 눈에 딱 들어왔고 순간 '이거다!'라고 생각했어요. 프랑스에서 배우고 익힌 기술을 접목시켜 실제 베이비 바나나같은 모양으로 만들어보기로 했어요. 바나나 모양의 실리콘 몰드를 직접 만들고, 코코아버터와 초콜릿을 섞은 노란색 베이스 소스를 만들어 디핑을 하고, 또 명암과 질감을 표현하기 위해 좀 더 진한 색 색소로 분사를 했어요. 그리고 더욱 더 실물과 비슷하게 보이기 위해 손으로 코코아파우더를 묻혀가며 세밀하게 작업했어요. 플레이팅을 할 때에는 굉장히 간단해 보이는 메뉴이지만 이것을 만들기 위해 3일이라는 시간이 소요되는, 장인 정신이 깃든 디저트랍니다.

사이드 메뉴로 커피 아이스크림을 곁들인 이유는 손님들이 프티 갸토와 커피를 함께 드시는 것에서 아이디어를 얻었어요. 달콤한 바나나 케이크를 먹은 후에 맛보는 시원한 커피 아이스크림은 palate cleanser(입 안을 깔끔하게 헹구는 디저트)의 역할을 하기도 하고요. 그리고 '어떻게 하면 더 손님들에게 재미있고 기억에 남게 다가갈 수 있을까?'라는 생각을 하며 프레젠테이션 방식에서 다시 한번 고민을 했어요. 제가 인간적으로도, 셰프로도 정말 좋아하고 존중하는, 당시 정식당 총괄 셰프였던 김호영 셰프님(현재는 뉴욕에서 너무 맛있고 핫한 레스토랑인 'JUA'의 오너 셰프)과 메뉴에 대한 많은 이야기를 항상 열정적으로 너무 재미있게 나누곤 했는데, 같이 이런저런 이야기를 하다가 '베이비 바나나를 과일 바구니에 담아서 소개하면 어떨까?'라는 아이디어가 나오게 되었어요. 그렇게 해서 과일 바구니에 신선하고 다양한 제철과일과 함께 담긴 베이비 바나나를 손님이 직접 집어 접시에 놓는, 정말 재미있는 서빙 방식이 탄생했죠.

The idea of serving coffee ice cream on the side came from customers enjoying coffee with petit gateau. Having a bite of cold coffee ice cream after eating a sweet banana cake works as a good palate cleanser. Also, I thought a lot about the presentation method and wanted to make it more interesting and memorable for the guests. I always used to have passionate and enjoyable conversations about menus with the executive chef of Jungsik, Chef Hoyoung Kim (currently the owner-chef of JUA, a very tasty and hot restaurant in New York), whom I respect not only as a chef but also as a person. One day, while we were having a conversation, he suggested an idea, 'How about serving the baby bananas in a fruit basket?' That's how we created an entertaining way of serving, where the customers pick up the bananas themselves from a fruit basket with a variety of fresh, seasonal fruits and place them on their plate.

나의 정체성을
표현한 디저트,
뉴욕-서울

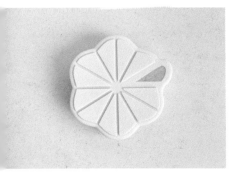

리제
LYSÉE

'뉴욕-서울'이라는 디저트는 코리안 프렌치 뉴요커의 아이덴티티를 가지고 있는 저와 비슷하게 프렌치 디저트인 슈를 바탕으로 한국의 볶은 현미를 이용한 크림과 캐러멜, 그리고 미국 사람들이 좋아하는 대표적인 견과류인 피칸과 아메리칸 스타일의 쿠키도우로 이루어져 있어요. 저처럼 세 가지의 문화가 공존하는 이 디저트는 피칸과 볶은 현미가 주는 고소한 맛과 부드럽고 비릿한 다양한 식감이 소화롭게 어우러져 개인적으로도 정말 좋아하는 디저트이고, 정식당에서도 메뉴에서 빠지지 않고 호불호 없이 오랫동안 사랑받은 대표적인 인기 메뉴예요. 저와 가장 닮아 있는 디저트라고 생각해 현재 저의 매장 리제에서도 '리제'라는 이름의 시그니처 메뉴로 선보이고 있어요.

디저트에서
가장 중요하게
생각하는 것

저는 일단 제철과일에 중심을 두고 메뉴를 만들어요. 봄, 여름, 가을, 겨울 계절마다 식재료가 모두 다르고, 제철에 먹는 과일이 단연 최고의 맛을 낸다고 생각하기 때문이에요. 만약 제철과일로 사과를 쓰기로 했다면, 사과에도 수많은 종류가 있기 때문에 원하는 맛을 가진 것을 찾기 위해 기본적으로 다양한 종류의 사과를 맛보고 결정한 후 여기에 어떤 맛을 추가해야 전체적으로 어울릴지를 고민해요. 그리고 손님들이 어떻게 하면 만든 사람의 의도를 쉽게 이해하면서 즐길 수 있을지, 어떤 텍스처가 좋을지, 제가 원하는 맛과 의도를 최대한 경험하실 수 있게끔 손님 입장에서 생각을 많이 하죠. 코리안 파인 다이닝이라는 레스토랑의 정체성이 있다 보니 한국적인 느낌을 접시에 담는 것에 대해서도 고민을 많이 하는 편이에요. 그리고 기본적인 맛, 텍스처, 디자인, 서비스뿐만 아니라 곁들여지는 다른 메뉴들과의 연결성, 정체성, 유니크함 등 정말 다양한 면에서 고민에 고민을 거듭하면서 만들어요. 물론 그중에서 가장 중요하다고 생각하고 중점을 두는 부분은 당연히 '맛'이에요.

뉴욕-서울
New York-Seoul

© Dan Ahn

Dessert that Expresses My Identity, New York-Seoul

Similar to myself, who has the identity of a Korean French New Yorker, the dessert 'New York-Seoul' is composed of French dessert choux as a base, with cream and caramel using Korean roasted brown rice, along with pecans which Americans love, and American style cookie dough. This dessert, which feels like three cultures coexist just like I am, is my personal favorite because it combines the savory taste of pecans and roasted brown rice with a variety of soft and crispy textures, and it is a popular menu that has been loved at Jungsik for a long time. I think it's the dessert that resembles me the most, so I'm currently offering it as a signature menu by the name 'LYSÉE' in my shop LYSÉE.

The Most Important Thing in Dessert

First, I work on the menu with a focus on seasonal fruits. It's because I believe that seasonal ingredients are different each spring, summer, fall, and winter, and the fruits in season definitely taste the best. For example, suppose I decide to use apple as the seasonal fruit. Because there are many types of apples, I always taste different apples to find the one with the taste I want and then decide which flavor to add to harmonize with the overall taste. And I think a lot from the customer's point of view so they can easily understand the intention of the person who made it, what kind of texture would work better, so they experience the taste and motive I portrayed as much as possible. Since the restaurant has a Korean Fine Dining identity, I am always concerned about adding a Korean touch to the dish. In addition to the fundamental factors- taste, texture, design, and service, I continue to contemplate while creating various aspects such as connectivity, identity, and uniqueness of the menu. But, of course, more than anything, 'taste' is the first and foremost factor I focus on.

© Dan Ahn

호롱
Horong

한국의 맛과 이미지를 디저트에 담다

조금 더 한국이라는 나라를 알리고 싶다는 마음에서 한국의 식재료를 사용한 한국적인 이미지의 디저트를 만들게 된 것 같아요. 음악, 드라마, 영화 등으로 한국의 문화가 세계에 널리 알려졌지만, 제가 프랑스로 갔던 2006년도만 해도 한국을 잘 모르는 사람들이 정말 많았었거든요. 그것에 관한 에피소드도 참 많고요. 그런 환경에 있다 보니 당연하게 한국이라는 나라를 더 많이 알리고 싶다는 생각이 들었던 것 같아요. 프랑스에 있는 동안 유학 시절부터 리옹에서 열리는 페이스트리 월드컵 pastry world cup을 매년 구경하러 갔었는데, 각국을 대표하는 파티시에 팀들이 이틀간 경쟁을 하는 세계적인 무대를 멀리서 지켜보며 꿈과 열정을 키우기도 했었죠. 한국도 좋은 식재료가 많은 나라고, 맛있는 디저트를 잘 만들 수 있는 세프가 있다는 것을 알리고 보여주고 싶어 프랑스에 있을 때에도 한국 식재료를 많이 사용하려 했고, 같이 일하는 셰프들에게도 경험하게 해주었어요. 또 항상 아이디어 노트에 이런 한국 식재료와 접목하면 좋겠다고 생각하는 아이디어를 적어두고, 한국에서 돌아다닐 때에도 이건 디저트로 멋지게 표현할 수 있겠다는 생각이 들면 모두 기록해두곤 했죠. 그런 노력과 열정, 아이디어들이 기와, 가마솥, 연꽃, 호롱 등의 다양한 디저트로 표현된 것이고요. 한국인이다 보니 제가 만드는 디저트에 한국적인 색이 느껴지고 들어가는 것은 특별한 것이 아니라 당연한 것 같아요. 그렇지만 또 그에 국한되고 싶지는 않고, 각 지역마다 나라마다 가지고 있는 여러 가지 다양한 식재료를 경험하고 다채롭게 응용하고 싶어요. 앞으로도 계속 새로운 식재료에 도전해보고 싶어요.

Embedding
the Taste and Image
of Korea into Dessert

I think I made desserts with Korean impressions using Korean ingredients, wanting to promote the country a little more. Korean culture is widely known around the world through music, drama, and movies, but when I went to France in 2006, there were so many people who didn't know Korea very well. And I have so many episodes about it, too. Being in such an environment, I think it was natural for me wanting to promote more about Korea. While I was in France, I went to see the Pastry World Cup held in Lyon every time ever since I was studying there, and I nurtured my dreams and passion by watching the world stage from a distance where representative patissier teams compete for two days. I wanted to inform and show that there are many interesting ingredients in Korea as well, and it's possible to make delicious desserts with them. I also let the chefs I worked with experience them as well. Also, I always wrote down any ideas that I thought would be nice to combine with these Korean ingredients in my idea notebook. Even when traveling around Korea, if I thought what I saw could be expressed beautifully as a dessert, I would write it all down. Such efforts, passion, and ideas were expressed in various desserts such as giwa (traditional Korean roof tiles), gamasot (Korean iron pot, cauldrons), lotus flowers, and lanterns. As I am a Korean, it seems that it's not unusual but rather goes without saying to give Korean identity to the desserts I make. However, I want to expand beyond it and experience and apply various ingredients each region has in each country. I want to continue to challenge using new ingredients in the course of time.

an Ahn

봄, 피어남
Bloom

파티시에를 꿈꾸는
이들에게

'디저트'라는 분야에도 제과, 제빵, 플레이팅 디저트, 초콜릿, 설탕 공예 등 다양한 카테고리가 있어요. 저는 일단 다양한 분야를 모두 경험해봐야 무엇을 더 잘하고 좋아하는지 알 수 있겠다 싶어 레스토랑, 페이스트리 부티크, 호텔, 베이커리 등 다양한 곳에서 경험을 했는데, 각각의 장단점이 있었어요. 우선 호텔은 개인적으로 정말 좋은 학교라고 생각해요. 저는 모든 디저트를 좋아하지만, 개인적 취향으로는 레스토랑의 플레이팅 디저트를 가장 좋아해요. 다른 곳들에 비해 큰 제약이 없다는 장점이 있기 때문이죠. 가지고 놀 수 있는 범위가 더 넓다고 해야 할까요? 뜨거운 과일

© VALRHONA USA

요리나 수플레부터 차가운 아이스크림까지 온도의 변화를 줄 수 있고, 식감이나 재료(특히 허브류)도 다채롭게 쓸 수 있고요. 아무래도 코스에서 가장 마지막에 나가는 메뉴이다 보니 테이크아웃이나 장기 보관 면에서 큰 장점이 있고, 손님들의 반응을 바로 접할 수 있다는 것도 좋은 부분이에요. 그렇지만 프레젠테이션, 사용하는 식기, 메뉴에 대한 설명 등 서비스에 대해서 많은 고민을 해야 한다는 것, 메뉴의 마지막을 맡고 있기에 어느 정도 앞의 메뉴와 밸런스나 컬러가 맞아야 한다는 것, 다른 분야의 팀들과도 협동해야 한다는 것, 그리고 주어진 시간에 손님들께 최고의 서비스를 빠르고 완벽하게 해야 한다는 것을 잊지 않아야 해요.

제빵을 배울 때 저는 반죽 만지는 촉감을 너무 좋아했었는데, 1년간 베이커리에서 일하면서 허리 디스크가 생겼어요. 물론 사람마다, 가게마다 환경이 다르고 제가 요령이 없었을 수도 있었겠지만, 밀가루 포대같은 무거운 것들을 수없이 들었다 나르고, 반죽을 할 때에도 큰 반죽 기계에 한꺼번에 많은 양을 투입하는 등 손목과 허리를 쓰는 일이 많아 육체적으로 쉽지 않았던 기억이 있어요. 시간대도 아침에 빵이 나와야 하니 밤이나 새벽에 출근해 아침에 들어오는, 일반 사람들과는 반대인 스케줄이 대부분이죠. 그러나 빵 냄새가 솔솔 풍기는 곳에서 아침을 맞이하고, 제가 손수 반죽한 것이 오븐에서 예쁜 빵이 되어 나올 때 무한한 희열을 느끼고, 갓 나온 빵을 먹을 때의 행복은 그 곳에서 일하는 사람만이 가질 수 있는 특권이라고 생각해요.

파티시에를 꿈꾸고 있는 분들께 제가 해주고 싶은 이야기는, 이 일이 쉽지 않은 일이기 때문에 힘들 때가 있더라도 본인의 꿈을 생각하며 일에 대한 열정과 사랑을

ALRHONA USA

To Those Who Dream of Becoming a Patissier

There are various categories within the field of 'desserts,' such as pastry, baking, plated desserts, chocolate, and sugar crafts. As for me, I wanted to know what I liked and was good at, so I tried all of them; restaurant, pastry boutique, hotel, and bakery, and each has its own pros and cons. First, I think hotels can be an excellent school based on my personal experience. I like all kinds of desserts, but I like the restaurant's plated desserts the most because it has the advantage of not being too restrictive compared to other fields. Should I say that there is a broader range to play with? From hot fruit dishes or souffle to cold ice cream, you can vary the temperature and use various textures and ingredients, especially herbs. Since it's the last menu to be served, it has great advantages in terms of taking them to-go or long-term storage, and you are able to check customers' reactions directly. However, keep in mind that you have to think a lot about servicing the desserts such as presentation, tableware, and explanations, that the balance or the color must match the previous menu to some extent since it takes responsibility for the last course of the meal, that you must cooperate and harmonize with teams in other sections, and that you should always lookout because you need to provide the best service to your guests quickly and perfectly at any given time.

During the baking course, I loved the texture of the dough in my hands, but after working at a bakery for a year, I was diagnosed with a herniated lumbar disc. Of course, each person and each shop had different situations or environments, and I might not have had the knack for it, but I had to carry many heavy objects such as bags of flour. Also, when kneading the dough, I had to transfer a large amount into large machines at once, and this was not easy physically because it would strain my wrist and back. Since the bread has to be served in the morning, most of the schedules are opposite of ordinary people, that we need to go to work at night or dawn and return in the morning. But, I believe it was my privilege to be able to greet the day in a place where you can smell the fresh-baked bread, feel the infinite joy of seeing the dough I made baked into beautiful bread in the oven, and the happiness of being able to eat the fresh-baked bread.

잃지 말라는 것, 그리고 할 수 있을 때 최대한으로 다양한 경험을 해보라는 것이에요. 겉보기에 아름다운 것을 만드는 직업이지만, 사실 그 이면에는 많은 체력과 정신력을 필요로 해요. 어떤 직업이든 힘들지 않은 직업이 없겠지만, 어렵고 포기하고 싶은 순간이 오더라도 처음에 시작한 그 마음과 열정을 떠올리면서, 자신의 꿈을 포기하지 말라고 이야기해주고 싶어요. 저는 워낙 먹는 것을 좋아하고 맛있는 음식을 먹으면 힐링이 되는 사람이라 여기저기 많이 먹으러 다니고, 새로운 문화를 받아들이는 것에 대해 항상 오픈되어 있어요. 또한 호기심도 많아 여행하는 것을 좋아하는데, 꼭 어딜 여행하면 그 곳의 새로운 식재료를 맛보고 경험하곤 해요. 때로는 그 경험을 위해 여행을 계획하기도 하고요. 아이디어 노트에 좋은 아이디어가 떠오르거나 흥미로운 것을 발견하면 항상 그리거나 적어두어요. 어떻게 보면 제가 즐기고 좋아하는 것들을 함과 동시에, 끊임없이 노력도 하고 있다고 생각해요. 지금도 변함없이 그렇게 하고 있고요. 저는 이 모든 경험들이 저에게 소중하고 큰 재산이라고 생각해요. 본인이 이루고자 하는 꿈, 열정과 사랑, 노력이 있으면 그게 무엇이든 언젠가는 다 됩니다. 파이팅!

What I want to say to those who dream of becoming a patissier is that this job is not easy. But even if there are difficult times, think of your dream and not lose your passion and love for it, and try to experience as many different things as possible while you can. What you see is a job that makes beautiful things, but it takes a lot of physical and mental strength. No profession is not challenging, but even if there comes a time when you feel like giving up, I want you to remember the desire and passion you had in the beginning and tell you not to give up on your dream. I am a person who loves to eat, and when I eat delicious food, I am content, so I travel to eat, and I am always open-minded when it comes to accepting new cultures. I also have lots of curiosity, so I like to travel, and whenever I do, I always taste and experience new ingredients. Sometimes I plan a trip for that experience. I always draw or take notes in my idea notebook when I come up with a good idea or find something interesting. So in a way, I think I am doing things that I enjoy and love, and at the same time, I am constantly striving and developing. But, even now, it's still the same. I consider all these experiences a precious and great asset to me. As long as you have a dream, passion, love, and effort for what you want to achieve, it will come true one day, whatever it is. Best of luck to you!

기와
Giwa

나의 첫 공간,
리제 LYSÉE

2022년 봄, 뉴욕에 오픈한 저의 첫 공간 '리제LYSÉE'는 하이엔드 디저트 숍이에요. 제 성인 'Lee'와 'Museum'이라는 뜻의 불어 'Musée(뮤제)'를 합친 'Lee's sweet museum'이라는 뜻이 담긴 이름이에요. 제가 오랫동안 꿈꿔 왔던 이름처럼 페이스트리 갤러리 같은 콘셉트인데, 저는 오래 전 파티시에가 되기로 결심한 이유도 그렇고, 항상 페이스트리를 '먹을 수 있는 예술 작품(edible art)'이라고 생각하거든요. 그게 제가 가진 직업의 가장 큰 매력이라고 생각하고요. 그런 점을 잘 살린 공간이에요. 코리안 프렌치 뉴요커라는 저의 문화적 정체성을 기반으로 세 가지 문화가 융화된 현대적이면서 클래식한, 뉴욕스러우면서도 한국적인 느낌이 공간과 메뉴에 녹아 있어요. 아름다우면서도 다채로운 디저트를 통해 동서양을 연결하는 문화적 매개체를 경험할 수 있는, 그런 의미 있고 멋진 공간을 생각하며 만들었어요. 기본적으로는 테이크아웃이 가능한 디저트 위주로 제공되는 부티크이지만, 다양한 구움 과자 뿐 아니라 하우스 메이드 드링크, 소량의 비에누아즈리도 제공되고 있고, 어느 정도의 시간이 흐른 뒤에는 제가 좋아하는 플레이팅 디저트 메뉴와 함께 디저트 테이스팅 코스를 시작하는 것이 목표예요. 많은 사랑과 응원 부탁드려요!

리제의 팀원 Matthieu Lobry, Annie Lee, Ashley Kelly, Olivia Moran

The team members at LYSÉE;
Matthieu Lobry, Annie Lee, Ashley Kelly, Olivia Moran

My First Space, LYSÉE

'LYSÉE,' a high-end dessert shop, is my first space that opened in New York in the spring of 2022. The name means 'Lee's Sweet Museum,' which is a combination of my last name 'Lee' and the French word 'Musée,' which means museum. The concept is like a pastry gallery, just like the name I've been dreaming of for a long time, and one of the reasons I decided to become a patissier is also because I always think of pastry as an 'edible art.' I think that's the biggest charm of my job, and it's a space that takes advantage of it. Based on my cultural identity as a Korean French New Yorker, a modern and classic blend of three cultures that has a New York-like and yet a Korean touch will be melted into the space and menu. I created it with such a meaningful and astonishing space in mind, where you can experience the cultural medium that connects East and West through beautiful and glamorous desserts. It is a boutique that mainly serves desserts you can take out, but we also provide various baked goods, house-made drinks, and a few viennoiseries. After some time, I aim to set up a dessert tasting course with my favorite plated dessert menus. Please show us a lot of love and support!

리제의 이야기가 실린 <뉴욕 타임스>
<The New York Time> with the story of LYSÉE

앞으로의 꿈

파티시에라는 꿈을 키운 열여덟에 '나는 내가 계획한 과정을 통해 10년 뒤에는 세계적인 파티시에가 되겠다.'라고 10년 계획표를 세웠듯이 저는 항상 단기적, 장기적인 꿈과 목표를 세워 두는 편이에요. 현재는 저의 첫 매장을 오픈하면서 30대의 목표로 계획했던 꿈을 이루어 나가는 중이에요.

저의 첫 가게 오픈을 준비하며 이미 여러 가지 어려움을 몸소 겪었어요. 어려울 거라고는 생각했지만 오픈이라는 것은 정말 보통일이 아니더군요. 그래서 가장 가까운 꿈과 목표는 '먼저 성공적으로 가게를 오픈하고, 좋은 팀을 만들고, 함께 성장하고, 계속해서 좋은 퀄리티를 유지하며 손님들에게 변치 않는 달콤한 행복감과 만족감을 주고 싶다.'라는 것이에요. 저는 손님들이 제 디저트를 통해 각박한 삶 속에서 조금이나마 위안을 얻었으면 좋겠고, 그런 디저트를 오랫동안 만들기 위해 항상 노력하는 사람이 되고 싶다는 마음으로 디저트를 만들어요. 이 마음을 언제나 변치 않고 간직하고 싶어요. 그 다음으로는 언젠가 뉴욕, 서울, 파리에 각각 제 디저트 공간을 하나씩 만드는 것이 꿈이에요. 그 후에는 아주 먼 미래가 될 수도 있겠지만, 언젠가 꼭 실력은 있지만 여건이 어려운 젊은 친구들이 디저트 셰프의 꿈을 이룰 수 있도록 서포트하며 돕고 싶습니다. 영감을 주고, 좋은 영향을 미치는 셰프로 기억되고 싶기도 하고요. 이 모든 것들을 가능하게 하기 위해서는 부단하게 열심히 노력해야 겠지요! 제가 너무나도 좋아하는, 그리고 앞으로도 여전히 좋아할 디저트 셰프로서의 일을 하면서 말이지요.

한국의 독자 분들에게

저의 첫 번째 책을 한국에서 출판하게 되어 저도 감회가 남다른데요. 언젠가 책을 내게 된다면 한국에서 먼저 출판하고 싶었고, 그 책에서 저의 디저트 철학과 스토리가 담긴 저만의 소중한 레시피, 그리고 제가 파티시에로서 걸어온 길을 이야기하고 싶었어요. 열여덟 제 꿈을 위해 한국을 떠나면서 가졌던 마음, 그동안 걸어왔던 길들과 다양한 경험들, 여러 가지 힘들었지만 재미있었고 행복했던 순간들, 에피소드 등을 독자 분들, 혹은 파티시에를 꿈꾸고 있는 분들과 공유하고 싶었고요. 디저트 셰프를 꿈꾸는 모든 분들에게 흥미롭고 희망적인 메시지가 되었으면 좋겠습니다.

Future Dream

I always set short-term and long-term dreams and goals, as I did at 18 when I developed my dream of becoming a patissier, 'I will become a world-class patissier in 10 years through the process I have planned.' Currently, I am in the process of fulfilling the dream I had planned for my 30s as I opened my first store.

While preparing to open my first store, I have already experienced several difficulties myself. I expected it would be difficult, but opening a shop was a different kind of challenge. So, my immediate dream and goal are to open my shop successfully, create a good team, grow together, maintain good quality, and provide customers happiness and satisfaction of sweetness that will not change. I make desserts with the intention that I want my customers to find some comfort in their harsh life through my desserts, and I want to be a person who always strives to make such desserts for a long time. With this in mind at all times, I wish to remain unchanged. My next dream is to create my own dessert shop in New York, Seoul, and Paris, one after another. After that, it may be a very distant future, but I would like to support and help young friends with skills yet under challenging circumstances to achieve their dreams of becoming pastry chefs. I want to be remembered as a chef who inspires and makes a positive impact. I will have to work hard persistently to make all these possible! Of course, while working as a patissier that I love and will continue to love.

For Korean Readers

I am deeply moved to publish my first book in Korea. If I were to publish my book someday, I first wanted it to be in Korea, and in it, I wanted to share my precious recipes with my dessert philosophy, stories, and the path I took as a patissier. I wanted to share my thoughts when I left Korea to pursue my dream at the age of 18, the paths I walked, numerous experiences, many challenges but fun and happy moments, and episodes with the readers or those who dream of becoming a patissier. I hope this book will be an exciting and hopeful message to all who dream of becoming a patissier.

1

/

Omija

/

오미자

저의 시그니처 프리 디저트Pre dessert이기도 한 '오미자'는 손님들께 'palate cleanser'로 많은 사랑을 받았던 '한입 디저트'입니다. 디저트에 사용할 수 있는 수많은 베리류의 과일들이 있지만, 몸에 좋아 한약재로도 쓰이고, 이름의 뜻처럼 다섯 가지 맛(단맛, 신맛, 쓴맛, 짠맛, 매운맛)을 느낄 수 있는 매력적이면서도 독특한 식재료인 오미자를 꼭 뉴욕에서 선보이고 싶었습니다. 뉴욕 현지의 식재료와 조화롭게 사용하고 싶어 여러 가지 허브를 찾던 중, 뉴욕의 어느 로컬 농장에서 재배한 교배종인 레몬바질이 새콤한 오미자와 잘 어울릴 것 같았고, 상큼함을 극대화시키기 위해 제가 가장 좋아하는 감귤류인 유자를 더해보았습니다. 유자 껍질에 진하게 배어 있는 유자의 향이 가열하기에는 너무 아까웠고, 아삭한 식감과 상큼하고 신선한 맛과 향을 어떻게 하면 가장 잘 살릴 수 있을지 고민하다가 김치를 담그듯 무침으로 만들게 되었습니다. 'palate cleanser'는 레스토랑의 코스 메뉴에서 마지막 메인(고기) 요리까지 마친 후 본격적인 디저트 메뉴로 들어가기 전, 입 안을 깔끔하게 헹구는 디저트를 의미합니다. 이런 의미로 '오미자'는 그 역할에 아주 충실한 디저트라고 생각합니다. 실제로 손님들께서 충분히 배가 부른 상태에서도 이 디저트를 먹는 순간, 입 안이 말끔하게 헹궈지고 개운해져 코스 요리를 처음부터 다시 시작할 수 있겠다고 우스갯소리를 하시기도 했고, 오미자의 이국적인 향과 매력에 즐거워 하시기도 했습니다. 한국의 식재료와 현지의 식재료가 잘 어우러진, 심플하지만 최고의 한입 디저트라고 생각합니다.

Ingredients

레몬바질 소르베

물	240g
우유	45g
라임 제스트	라임 2개 분량
레몬그라스	1/2개
레몬바질	6g
설탕	78g
글루코스 (파우더)	10g
아이스크림 안정제 (stabilizer)	1g
라임즙	30g

오미자 주스

물	270g
설탕	45g
건조 오미자	18g

Directions

1. 냄비에 물, 우유, 라임 제스트, 레몬그라스를 넣고 끓어오를 때까지 가열한다.
2. 불을 끈 후 레몬바질을 넣고 볼 입구에 랩을 씌워 10분간 인퓨징한다.
3. 체에 거른다.
4. 체에 거른 **3**을 40°C까지 가열한 후 미리 섞어둔 설탕, 글루코스, 아이스크림 안정제를 넣고 1분 정도 끓인다.
5. 불에서 내려 충분히 식혀 라임즙을 넣고 블렌더로 믹싱한 후 하루 동안 냉장고에서 숙성시킨다. 사용하기 전 한 번 더 블렌더로 믹싱해 아이스크림 머신에서 소르베로 만든다.

   ● 소르베 베이스를 제대로 식히지 않은 상태에서 라임즙을 섞게 되면 우유의 유지방이 산에 의해 분리되므로 충분히 식힌 후 라임즙을 섞는 것이 좋다.
   아이스크림 머신 대신 파코젯Pacojet을 사용해도 좋다.

1. 냄비에 물 1/3, 설탕을 넣고 설탕이 녹을 때까지 가열해 불에서 내린다.
2. 건조 오미자를 넣고 잘 섞은 후 남은 물을 더한다.
3. 냉장고에서 하루 동안 인퓨징한 후 다음 날 체에 걸러 사용한다.

Lemon Basil Sorbet

240 g	Water
45 g	Milk
2	Limes, zested
1/2 stalk	Lemongrass
6 g	Lemon basil
78 g	Sugar
10 g	Glucose (powder)
1 g	Ice cream stabilizer
30 g	Lime juice

1. Heat water, milk, lime zest, and lemongrass in a saucepan until it boils.
2. Remove from the heat, add lemon basil and cover with plastic wrap, and infuse for 10 minutes.
3. Filter through a sieve.
4. Heat sieved **3** to 40°C, then stir in previously mixed sugar, glucose, and ice cream stabilizer. Boil for about 1 minute.
5. Remove from the heat and cool sufficiently. Add lime juice, mix using a blender and refrigerate for a day. Mix with a blender once more before use, then make it into sorbet using the ice cream maker.

   ● If you mix lime juice without cooling the sorbet base properly, the milk fat in the milk will separate by acid, so it is better to mix the lime juice after cooling the sorbet base sufficiently. If you don't have an ice cream maker, you can also use Pacojet.

Omija Juice

270 g	Water
45 g	Sugar
18 g	Dried omija

1. In a saucepan, heat 1/3 of water and sugar just until the sugar dissolves. Remove from the heat.
2. Stir in dried omija and add the remaining water.
3. Refrigerate overnight to infuse, and sieve the next day to use.

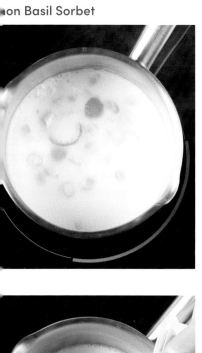

2

3

5

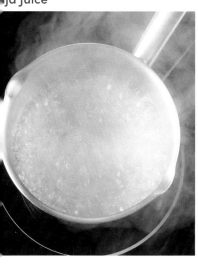

2

3

유자 무침❖

유자 껍질	55g
유자즙	15g
설탕	25g
꿀	5g

1. 볼에 잘게 채 썬 유자 껍질, 유자즙, 설탕, 꿀을 넣고 골고루 버무린다.
2. 볼 입구를 랩핑해 실온에서 3일간 둔 후 냉장고에 보관한다.

유자 샐러드

깍게 지른 사과	55g
유자 무침❖	20g
유자청	5g
유자 제스트	적당량

볼에 0.5cm 크기로 깍뚝썰기한 사과, 유자 무침, 유자청, 유자 제스트를 넣고
골고루 섞는다.

● 유자 샐러드는 오래 보관하면 사과에서 물이 생겨 물러지기 때문에 서비스 직전에 만들어 바로
사용하는 것이 좋다. 사과의 품종은 홍옥 중에서도 아삭하면서 과즙이 풍부하며 약간의 산미가 있는
것을 선택하는 것이 좋다. 뉴욕에서는 로컬 사과를 사용했다.

Yuja Muchim❖

55 g	Yuja peel
15 g	Yuja juice
25 g	Sugar
5 g	Honey

1. In a bowl, add thinly sliced yuja peels, yuja juice, sugar, and honey. Mix with your hands thoroughly.
2. Wrap the bowl and keep it at room temperature for 3 days, then store in a refrigerator.

* *Muchim*: A dish in which various seasonings are added to the ingredients, usually vegetables, and mixed well with hands so that the seasoning is well absorbed.

Yuja Salad

55 g	Apple, diced small
20 g	Yuja muchim❖
5 g	Yuja preserve
QS	Zest of yuja

In a bowl, add apples cut into 0.5 cm size cubes, yuja muchim, yuja preserve, yuja zest, and toss to coat evenly.

● If you store the salad for a long time, moisture will extract from the apple and become soggy. Therefore it is best to make it just before serving. It's better to choose the apple from the crunchy variety among Jonathans and has a bit of acidity. In New York, I used local apples.

Muchim

2

Salad

1

2

| **플레이팅** | **1.** | 작은 볼 정중앙에 유자 샐러드를 한 스푼 올린 후 레몬바질 소르베를 한 스쿱 올리고 신선한 허브로 장식한다. 오미자 주스는 소스 컵에 담아 따로 준비한다. |
| | **2.** | 서비스 시 오미자 주스를 레몬바질 소르베 왼쪽으로 조심스럽게 부어준다. |

| Plating | **1.** | Put a spoonful of yuja salad in the center of a small bowl, place a scoop of lemon-basil sorbet, and decorate with fresh herb. Prepare the omija juice in a separate sauce cup. |
| | **2.** | When serving, carefully pour the omija juice to the left of the lemon basil sorbet. |

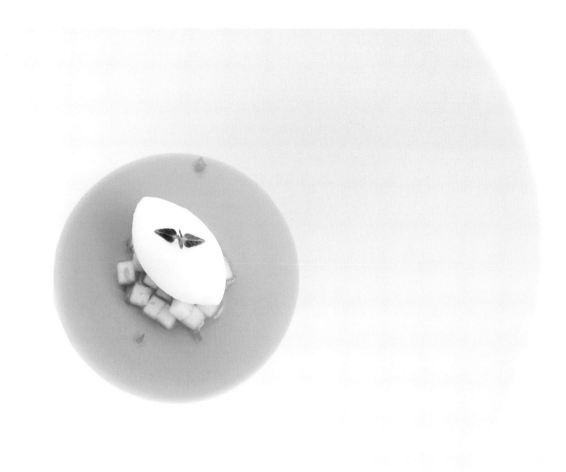

'Omija,' which is also my signature pre-dessert, is a bite-sized dessert adored by customers as a 'palate cleanser.' There are many types of berries that can be used in desserts, but I wanted to introduce *omija** in New York. This attractive and unique ingredient is good for the body and used as herbal medicine and has five different flavors (sweet, sour, bitter, salty, spicy), as the name implies. I wanted to use it in harmony with local ingredients in New York, so I looked for a variety of herbs. Among them, I tried lemon basil, a hybrid grown on a local farm in New York, because I thought it would go well with the sour omija. And to maximize the tanginess, I added *yuja**, my favorite citrus fruit. The aroma of yuja, which is deeply permeated in the peel, was too precious to be heated. Consequently, it got me thinking about how to emphasize best the crunchy texture and the refreshing, fresh taste and scent, and I seasoned it as if to make kimchi. A 'palate cleanser' refers to a dessert that refreshes your mouth clean before engaging the full-fledged dessert menu after finishing the final main (meat) dish in a restaurant's course menu. In that sense, I think 'omija' is a dessert that is very faithful to its role. In fact, even when the guests were sufficiently full, the moment they ate this dessert, their mouths were thoroughly cleansed and refreshed; they would playfully say that they could start the course dishes from the beginning again. I believe it's a menu that is the best bite-sized dessert even though it is simple and a harmonious combination of Korean ingredients and local ingredients.

* *Omija* translates to a 'fruit with five flavors' and is also known as Schisandra.
* *Yuja* is the Korean name for yuzu, a type of citron.

2

/

Sujeonggwa -Pear Tart

/

수정과 배 타르트

디저트를 만들 때 가장 중요하게 생각하는 것 중 하나가 '제철 과일을 사용하는 것'입니다. 제철에 먹는 과일이 가장 맛있다는 것이 저의 철학이고, 개인적으로 과일이 많이 들어간 디저트를 좋아하기 때문이기도 합니다. 저는 과일 타르트를 유독 좋아해서 항상 제철 과일을 사용한 타르트를 메뉴에 넣곤 하는데요, 수정과 배 타르트는 향신료의 계절인 가을에 선보였던 디저트 중 하나 입니다.

프랑스나 미국에서 디저트 재료로 많이 사용되는 서양 배와는 달리 한국 배(아시안 페어Asian pear라고도 불립니다.)는 디저트 의 재료로 사용되는 것을 본 적이 없습니다. 프랑스에서는 한국 배가 아주 생소한 과일이기도 했고요. 생소한 재료, 익숙하지 않 은 재료로 도전해보는 것을 좋아하는 저는 한국 배를 사용한 디저트를 만들어보기로 했습니다. 한국 배 특유의 아삭한 식감과 풍 부한 수분감을 최대한 살리는 것이 좋다고 판단했고, 배 자체의 풍미를 더 강하게 끌어올리기 위해 한국 배와 서양 배를 함께 사 용했습니다. 그래서 한국 배와 조화롭게 어울릴 수 있는 서양 배를 찾기 위해 모든 품종의 배를 하나씩 분석해보고 코미스 페어 comice pear라는 품종을 사용하기로 결정했습니다. 특유의 꽃 향과 풍미, 잘 익었을 때 입안에서 느껴지는 사르르 녹는 듯한 질감과 과즙이 특징인 코미스 페어는 프랑스에 있을 때부터 제가 가장 좋아했던 품종입니다. 모든 마켓에서 판매하는 구하기 쉬운 품종 은 아니어서 출근 전 코미스 페어를 파는 마켓에 들러 직접 보고 가장 맛이 좋을 것 같은 것으로 수십 개씩 골라 양손 가득 봉지를 들고 땀을 뻘뻘 흘리며 출근했던 기억이 있습니다.

프랑스에서는 가을과 겨울의 음료로 뱅쇼를, 미국에서는 스파이스 라테를 떠올리듯 한국의 가을과 겨울 음료로는 수정과가 가장 먼저 생각났습니다. (물론 쌍화차도 떠오르긴 하지만 수정과가 배와 더 잘 어울린다고 판단했습니다. 한국의 '을지다방'이라 는 곳에서 쌍화차를 맛보고 영감을 얻었는데, 언젠가 그 기억을 살려 쌍화차를 이용한 저만의 방식으로 풀어낸 디저트도 만들어 보고 싶습니다.) 가을이라는 계절, 후식이라는 테마, 배라는 재료가 함께 잘 어우러질 수 있도록 전통적인 맛보다는 조금 덜 달게, 덜 진하게 만들어 타르트와 함께 먹었을 때 하나의 맛이 도드라지지 않고 전체적으로 조화롭게 느껴지도록 만들었습니다.

이 디저트에서 주인공은 단연 '배'이고 단맛은 배 하나로도 충분했기에 설탕은 거의 사용하지 않았습니다. 조리한 서양 배, 유 자 시럽에 담근 서양 배, 한국 배, 밤꿀 시럽에 절인 한국 배, 한국 배 소르베까지 다양한 조리 방법을 활용해 풀어냈기에 다양한 배의 매력을 맛보는 재미를 느낄 수 있는 디저트라고 생각합니다. 한국과 서양이 자연스럽게 연결되는, 상큼하면서도 은은한 가 을의 향을 만끽할 수 있는 디저트입니다.

Ingredients

10인분 ◆ SERVES 10

아몬드 크림

버터	50g
설탕	50g
아몬드가루	50g
달걀	50g
다크 럼	5g

타르트

버터	70g
슈거파우더	45g
소금	한꼬집
아몬드가루	10g
달걀	27g
중력분	115g
서양 배	적당량

Almond Cream

50 g	Butter
50 g	Sugar
50 g	Almond powder
50 g	Eggs
5 g	Dark rum

Tart

70 g	Butter
45 g	Powdered sugar
Pinch of Salt	
10 g	Almond powder
27 g	Eggs
115 g	All-purpose flour
QS	Western pears

Directions

1. 볼에 깍둑썰기한 차가운 상태의 버터, 설탕을 넣고 재료가 골고루 섞일 때까지 믹싱한다.
2. 아몬드가루를 넣고 골고루 섞일 때까지 믹싱한다.
3. 미리 섞어둔 달걀과 다크 럼을 조금씩 넣어가며 믹싱한다.
4. 모든 재료가 골고루 섞이면 짤주머니에 담아 냉장고에 보관한다.

1. 볼에 버터, 슈거파우더, 소금을 넣고 재료가 골고루 섞일 때까지 믹싱한다.
2. 아몬드가루를 넣고 골고루 섞일 때까지 믹싱한다.
3. 달걀을 조금씩 넣어가며 믹싱한다.
4. 체 친 중력분을 넣고 반죽이 한 덩어리가 될 때까지 믹싱한다.
5. 반죽 위아래에 테프론시트를 깔고 1.5mm 두께로 밀어 편 후 냉장고에 잠시 보관한다.
6. 지름 9cm 원형 틀로 반죽을 자른다.

1. Beat cold butter cut into cubes and sugar in a bowl until evenly mixed.
2. Add almond powder and mix thoroughly.
3. Combine previously mixed eggs and dark rum a little bit at a time.
4. Once all the ingredients are combined, put them in a piping bag and refrigerate.

1. Beat butter, powdered sugar, and salt in a bowl until evenly mixed.
2. Add almond powder and mix thoroughly.
3. Combine eggs little by little.
4. Add sifted all-purpose flour and mix until the dough forms a ball.
5. Place the dough between two Teflon sheets and roll out to 1.5 mm thickness. Store in a refrigerator for a while.
6. Cut the dough with a 9 cm round ring.

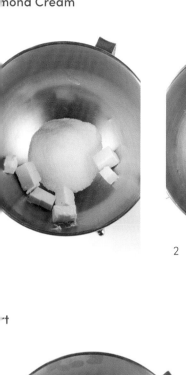

2

3

4

t

2

3

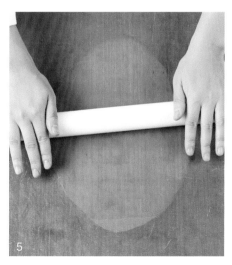

5

6

7. 지름 6cm 타르트 링에 자른 반죽을 넣고 퐁사주한다.

8. 나이프로 여분의 반죽을 깔끔하게 정리한다.

9. 스파이크 롤러나 포크를 이용해 반죽 바닥에 구멍을 낸다.

10. 타공 매트에 올려 167°C로 예열된 오븐에서 6~7분간 구운 후 타르트 링을 제거해 식힌다.

11. 식은 타르트에 아몬드 크림을 7g씩 파이핑한다.

12. 숟가락이나 미니 스패출러로 아몬드 크림을 평평하게 정리한다.

13. 서양 배를 얇게 썬다.

14. 얇게 썬 서양 배를 아몬드 크림 위에 올린 후 가볍게 눌러준다.

15. 167°C로 예열된 오븐에서 6~7분간 굽는다.

7. Fonçage the dough in 6 cm diameter tart rings.

8. Trim off the excess dough clean with a paring knife.

9. Prick holes in the dough with a spike roller or a fork.

10. Place over a perforated mat and bake in an oven preheated to 167°C for 6~7 minutes. Remove the tart rings after baking and let cool.

11. Pipe 7 grams of almond cream in the cooled tart.

12. Spread the almond cream evenly using a spoon or a mini spatula.

13. Thinly slice the Western pears.

14. Place thinly sliced pears over almond cream and press lightly.

15. Bake in an oven preheated to 167°C for 6~7 minutes.

8

9

11

12

14

15

무설탕 커스터드 크림✤

우유	113g
생크림(36%)	12g
바닐라빈	1/4개
노른자	20g
옥수수전분	6g
박력분	6g
젤라틴매스(200bloom)	10g
버터	14g

1. 냄비에 우유, 생크림, 바닐라빈을 넣고 끓어오를 때까지 가열한다.
   ● 우유의 일부는 노른자와 섞기 위해 남겨둔다.
2. 볼에 노른자, 남겨둔 우유를 넣고 휘퍼로 가볍게 섞어준 후 1을 넣어가며 섞는다.
3. 다시 냄비에 옮겨 체 친 옥수수전분, 박력분을 넣고 휘퍼로 재빠르게 섞어가며 호화시킨다.
4. 충분히 호화되면 불에서 내려 젤라틴매스, 버터를 넣고 섞는다.
5. 바믹서로 유화시킨 후 급속 냉동고나 냉동실에 잠시 두어 온도를 4°C까지 낮춰 냉장고에 보관한다.
   ● 급속 냉동고나 냉동실에 잠시 두는 것은 급격하게 온도를 떨어뜨려 살균하기 위함이다.

꿀 크림

크렘 프레슈	20g
꿀	15g
무설탕 커스터드 크림✤	75g

1. 볼에 크렘 프레슈를 넣고 가볍게 풀어준다.
   ● 크렘 프레슈는 프랑스 유제품 중 하나로 유지방 함량이 45% 정도로 점도가 되직한 편이며 산미가 느껴지는 것이 특징이다. 크렘 프레슈 대신 플레인 요거트로 대체 가능하다.
2. 다른 볼에 꿀, 무설탕 커스터드 크림을 넣고 잘 섞는다.
3. 1과 2를 섞은 후 짤주머니에 넣어 냉장고에 보관한다.

Sugar-free Custard Cream ✤

113 g	Milk
12 g	Heavy cream (36%)
1/4	Vanilla bean
20 g	Egg yolks
6 g	Corn starch
6 g	Cake flour
10 g	Gelatin mass (200bloom)
14 g	Butter

1. Heat milk, heavy cream, and vanilla bean in a pot until it boils.
   ● Set aside a portion of the milk to combine with egg yolks.
2. Lightly mix egg yolks and remaining milk with a whisk. Continue mixing while gradually adding 1.
3. Transfer back to the pot, add sifted corn starch and cake flour and cook to gelatinize while stirring quickly with a whisk.
4. Once it's gelatinized, remove from the heat and stir in gelatin mass and butter.
5. Emulsify using a hand blender, then keep in a blast freezer or a freezer for a while to drop the temperature to 4°C, then store in a refrigerator.
   ● Keeping in a blast freezer or a freezer for a while is to sterilize by rapidly dropping the temperature.

Honey Cream

20 g	Crème fraîche
15 g	Honey
75 g	Sugar-free custard cream✤

1. Lightly loosen crème fraîche in a bowl.
   ● Crème fraîche is one of the French dairy products, with a milk fat content of 45%, relatively thick, and characterized by an acidic taste. You can substitute it with plain yogurt instead of crème fraîche.
2. Mix the sugar-free custard cream and honey thoroughly in a separate bowl.
3. Combine 1 and 2, put in a piping bag, and store in a refrigerator.

ar-free Custard Cream

2

3

5

ey Cream

2

3

수정과

물	300g
생강	8g
시나몬스틱	6g
흑설탕	10g
황설탕	12g
밤꿀	10g

냄비에 모든 재료를 넣고 강불로 가열하기 시작하다가 끓어오르면 중불로 낮춰 25~30분 정도 졸인 후 체에 걸러 냉장고에 보관한다.

무설탕 배 콩포트

깍둑썰기한 서양 배	100g
레몬즙	약간
바닐라파우더	0.3g

1. 냄비에 작게 깍둑썰기한 서양 배, 레몬즙을 넣고 수분이 날아갈 때까지 저어가며 가열한다.

2. 서양 배에서 더 이상 수분이 나오지 않고 걸쭉해진 상태가 될 때까지 졸인다.

3. 불에서 내려 바닐라파우더를 넣고 섞은 후 냉장고에 보관한다.

Sujeonggwa

300 g	Water
8 g	Ginger
6 g	Cinnamon stick
10 g	Dark brown sugar
12 g	Light brown sugar
10 g	Chestnut honey

Boil all the ingredients in a saucepan over high heat. Once it boils, reduce the heat to medium and simmer for 25~30 minutes. Strain and refrigerate.

Sugar-free Pear Compote

100 g	Western pear, diced
	Dash of Lemon juice
0.3 g	Vanilla powder

1. Put the pear diced small and lemon juice in a pot and heat while stirring until moisture evaporates.

2. Simmer the pear until it becomes thick and no more moisture comes out of the pear.

3. Remove from the heat, stir in vanilla powder, and store in a refrigerator.

ar-free Pear Compote

2

3

배 소르베

물	100g
꿀	6g
백설탕	35g
황설탕	18g
아이스크림 안정제 (stabilizer)	0.5g
배 퓌레	165g
한국 배즙	115g

1. 냄비에 물, 꿀, 백설탕, 황설탕, 아이스크림 안정제를 넣고 섞은 후 끓인다.
2. 배 퓌레가 담긴 볼에 1을 붓고 잘 섞는다.
3. 한국 배즙을 넣고 블렌더로 섞는다.
4. 냉장고에서 하루 동안 숙성시킨 후 아이스크림 머신에서 소르베로 만든다.
   ● 아이스크림 머신 대신 파코젯Pacojet을 사용해도 좋다.

유자 시럽❖

유자즙	65g
물	80g
유자청	20g

냄비에 모든 재료를 넣고 끓어오를 때까지 가열한 후 냉장고에 보관한다.

서양 배

서양 배	1개 분량
유자 시럽❖	전량

서양 배를 5~6cm 길이의 삼각형 모양으로 일정하게 자른 후 유자 시럽에 담가 유자 향이 충분히 배어들도록 한다.

Pear Sorbet

100 g	Water
6 g	Honey
35 g	Sugar
18 g	Light brown sugar
0.5 g	Ice cream stabilizer
165 g	Pear puree
115 g	Korean pear juice

1. Combine water, honey, sugar, light brown sugar, and ice cream stabilizer in a saucepan and boil.
2. Pour 1 into a bowl of pear puree and mix.
3. Add Korean pear juice and combine with a blender.
4. Refrigerate for a day and make sorbet using the ice cream maker.
   ● If you don't have an ice cream maker, you can also use Pacojet.

Yuja Syrup❖

65 g	Yuja juice
80 g	Water
20 g	Yuja preserve

Add all the ingredients to a saucepan and bring to a boil. Store in a refrigerator.

Western Pear

1	Western pear
All of Yuja syrup❖	

Slice Western pear into triangles of 5~6 cm length and store in yuja syrup so the yuja aroma can permeate sufficiently.

2

3

Syrup

밤꿀 시럽❖

물	150g
밤꿀	27g
흑설탕	5g
레몬 제스트	레몬 1개 분량
오렌지 제스트	오렌지 1/2개 분량

1. 냄비에 물, 밤꿀, 흑설탕을 넣고 흑설탕이 모두 녹을 때까지 끓인다.

2. 레몬 제스트, 오렌지 제스트를 넣고 끓인다.

3. 끓어오르면 불에서 내린 후 볼 입구를 랩핑해 냉장고에 보관한다.

한국 배

한국 배	1개 분량
밤꿀 시럽❖	전량

한국 배를 지름 0.5cm 정도의 작은 구형으로 파내어 밤꿀 시럽과 함께 진공 머신(수비드 머신)으로 공기가 들어가지 않도록 압축해 절인다.

● 진공 머신이 없는 경우 밤꿀 시럽에 구형으로 자른 한국 배를 담가 두어도 좋다. 한 접시당 4개 정도가 사용되므로 약 40개를 준비한다.

● 플레이팅에 사용할 큰 구형으로 파낸 한국 배도 준비한다. 한 접시당 2개 정도가 사용되므로 약 20개를 준비한다.

Chestnut Honey Syrup❖

150 g	Water
27 g	Chestnut honey
5 g	Dark brown sugar
Zest of 1 Lemon	
Zest of 1/2 Orange	

1. Heat water, chestnut honey, and dark brown sugar in a saucepan; boil until all the dark brown sugar dissolves.

2. Add lemon zest and orange zest, and continue to boil.

3. Once it boils, remove from the heat, and cover the bowl with plastic wrap, and refrigerate.

Korean Pear

1	Korean pear
All of Chestnut syrup❖	

Scoop out Korean pear into small spheres, about 0.5 cm in diameter. Marinate the pears in chestnut honey syrup by compressing them together using a vacuum machine (sous vide machine) to prevent air from getting in.

● If the vacuum machine is unavailable, you can soak the pear spheres in the chestnut honey syrup. Prepare about 40 pieces since around four are used per plate.

● Prepare Korean pear scoped out into large spheres to be used for plating. Prepare about 20 pieces since about two are used per plate.

estnut Honey Syrup

 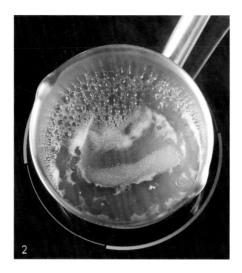 

ean Pear

플레이팅

1. 접시 왼쪽에 타르트를 올린 후 그 옆에 꽃 모양으로 잘라낸 배 한 조각을 올려준다.

2. 타르트에 꿀 크림을 파이핑한다.

3. 남은 공간에 무설탕 배 콩포트를 채운다.

4. 무설탕 배 콩포트 위에 유자 시럽에 담가둔 삼각형 모양의 서양 배 2조각을 올린다.

5. 큰 구형의 한국 배, 밤꿀 시럽에 절인 작은 구형의 한국 배로 장식한다.

6. 레드 소렐을 올린다.

7. 티뭇후추를 뿌린다.

8. 꽃 모양 배 위에 배 소르베를 올린다.

9. 수정과는 따로 잔에 담아 테이블에서 서비스한다.

8

9

| Plating | 1. | Place a tart on the left of the plate and a piece of pear cut into a flower shape next to it. |

Plating

1. Place a tart on the left of the plate and a piece of pear cut into a flower shape next to it.
2. Pipe honey cream on the tart.
3. Fill the remaining space with the sugar-free pear compote.
4. Place two pieces of triangle-shaped Western pear soaked in yuja syrup over the sugar-free pear compote.
5. Decorate with large spheres of Korean pear, then with small spheres of Korean pear soaked in chestnut honey syrup.
6. Put red sorrel leaves.
7. Sprinkle timut pepper.
8. Place pear sorbet over the flower-shaped pear.
9. Provide sujeonggwa in a separate cup and serve at the table.

One of the most important things to consider when making desserts is to use seasonal fruits. It is partly because of my philosophy that seasonal fruits taste the best and partly because I personally like desserts with a lot of fruit. I especially like fruit tarts, so I always put tarts made with seasonal fruits on the menu. The *Sujeonggwa**-Pear tart is one of the desserts presented in autumn, the season of spices.

Unlike Western pears, often used as a dessert ingredient in France or the United States, I have never seen Korean pears (also called Asian pears) used as a dessert ingredient. In France, Korean pears were also a very unfamiliar fruit. I like to experiment with unaccustomed and unusual ingredients, so I decided to try making a dessert using Korean pears. I decided that it would be good to maximize the crisp texture and rich moisture unique to Korean pears, and I chose to use Korean and Western pears together to enhance the flavor of the pears themselves. So, I went on to buy and analyze all kinds of pears one by one to find the Western pear that can pair well with Korean pear and decided to use a variety called comice pear. With its distinctive floral scent and flavor, the texture that melts in your mouth with its nectar when it's ripe, comice pear has been my favorite variety since I was in France. Because it wasn't sold in all markets, it was not an easy kind to find. I remember stopping by the market selling comice pear on the way to work and hand-picked dozens that I thought would taste the best, and going to work carrying bags full of the fruit in both hands, sweating profusely.

Just as vin chaud is enjoyed in autumn and winter in France, and spice latte in the States, sujeonggwa definitely came to mind as a drink in Korea for autumn and winter. (Of course, *ssanghwacha** also comes to mind, but I decided that sujeonggwa suited pear better. I was inspired by tasting ssanghwacha at a place called 'Eulji Dabang (Eulji coffee shop)' in Korea; I would like to make use of the memory and create a dessert my own way someday using ssanghwacha.) To harmonize the season called autumn, the theme of dessert, and the ingredient of pear, I made it a little less sweet and less intense than the traditional flavors so they could harmonize well, without any of the flavor standing out, with the tart.

In this dessert, the lead role is definitely the pear, and the pear itself was sweet enough, so I barely used sugar. I think it is a dessert you can enjoy the various charm of pears which use different cooking methods- including cooked Western pear, Western pear soaked in yuja syrup, Korean pear, and Korean pear pressed in chestnut honey syrup, and Korean pear sorbet. It's a dessert where you can enjoy the fresh yet subtle scent of autumn, where Korea and the West are connected naturally.

* *Sujeonggwa* is a Korean traditional cinnamon punch made with dried persimmon and ginger.
* *Ssanghwacha* is a traditional Korean medicinal tea made with several medicinal herbs.

3

/

Raspberry Vacherin

/

라즈베리 바슈랭

'라즈베리 바슈랭'은 2021년에 밸런타인데이를 기념해 특별히 만든 디저트입니다. '바슈랭'이라는 디저트는 본래 아이스크림이나 소르베, 머랭이 주가 되는 프렌치 디저트인데, 이 디저트를 밸런타인데이 콘셉트에 맞춰 붉은 장미 모양으로 만들고 싶어 강렬하면서도 예쁜 붉은 색을 내기 위해 라즈베리를 사용했고, 개인적으로 가장 좋아하는 향신료인 티뭇후추를 곁들였습니다. 티뭇후추는 동남아시아 네팔이 원산지인데 특유의 꽃 향이 강하면서 향기롭고, 후추인데도 상큼한 느낌이 있어 베리류나 시트러스류 과일과도 아주 잘 어울리는 향신료입니다. 프랑스에 있었을 당시 플레인 요거트나 프로마주 블랑에 바닐라 설탕이나 꿀을 살짝 뿌리고 신선한 라즈베리를 가득 얹어 먹는 것을 좋아했었는데, 이와 비슷한 느낌으로 크렘 프레슈를 함께 곁들여주었습니다. 마지막 포인트인 머랭은 잎사귀 모양으로 만들어 장미를 감싸는 듯한 느낌을 주었습니다. 밸런타인데이라는 주제에 맞게 아주 강렬한 붉은색이 포인트인, 그림처럼 매력적인 디저트입니다.

Ingredients

10인분 ◆ SERVES 10

라즈베리 소르베

냉동 라즈베리	300g
설탕A	50g
라벤더 꿀	5g
라즈베리즙	75g
글루코스(파우더)	10g
설탕B	10g

라즈베리 주스

냉동 라즈베리	300g
설탕	45g

Directions

1. 볼에 냉동 라즈베리, 설탕A, 라벤더 꿀을 넣고 섞어 하루 동안 냉장고에서 숙성시킨다.

2. 냄비에 라즈베리즙을 넣고 40°C까지 가열한 후 미리 섞어둔 글루코스와 설탕B를 넣고 1분간 끓인다.

3. 1과 2를 섞은 후 체에 걸러 식힌다.

4. 블렌더로 유화시킨 후 아이스크림 머신에서 소르베를 만든다.
   - 아이스크림 머신 대신 파코젯Pacojet을 사용해도 좋다.

5. 장미 모양 실리콘 틀에 라즈베리 소르베를 채운다.
   - 여기에서는 아마존 사이트에서 구매한 장미 모양 실리콘 틀을 사용했으나, 시판 틀도 그 크기와 모양이 어울리는 것을 사용해도 좋다.

6. 미니 스패출러로 빈틈이 없도록 정리한다.

7. 냉동실에서 완전히 얼린 후 실리콘 틀에서 빼 사용한다.

1. 볼에 냉동 라즈베리, 설탕을 넣고 섞은 후 볼 입구에 랩을 씌운다.

2. 뜨거운 물이 담긴 볼 위에 받쳐 30분 정도 중탕한다.

3. 주스가 만들어지면 면포에 걸러 과육을 분리한다.

Raspberry Sorbet

300 g	Frozen raspberries
50 g	Sugar A
5 g	Lavender honey
75 g	Raspberry juice
10g	Glucose (powder)
10 g	Sugar B

1. Combine frozen raspberries, sugar A, and lavender honey, and let the mixture rest in a refrigerator for a day.

2. Heat raspberry juice in a pot to 40°C, stir in previously mixed glucose and sugar B, and boil for one minute.

3. Combine 1 and 2, strain the mixture and let cool.

4. Emulsify with a blender, and make sorbet using the ice cream maker.
   - If you don't have an ice cream maker, you can also use Pacojet.

5. Fill the rose-shaped silicon mold with the sorbet.
   - Here, I used a rose-shaped silicon mold bought from Amazon. You can use a commercially available mold of similar size and shape.

6. Use a mini spatula to fill thoroughly so there are no gaps.

7. Freeze completely in a freezer, and remove from the mold to use.

Raspberry Juice

300 g	Frozen raspberries
45 g	Sugar

1. Combine frozen raspberries and sugar in a bowl, and cover with plastic wrap.

2. Heat over a double boiler for about 30 minutes.

3. When the juice is made, filter it through a cloth to separate it from the pulp.

pberry Sorbet

 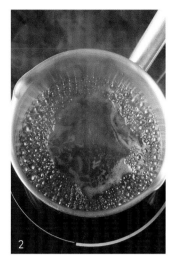  

2 3 4

6 7

pberry Juice

2 3

후추 무스

생크림(36%)	380g
티뭇후추	2g
젤라틴매스(200bloom)	10g
화이트초콜릿	88g
(발로나 IVOIRE 35%)	

1. 냄비에 생크림 1/3, 티뭇후추를 넣고 끓어오를 때까지 가열한 후 불에서 내려 10분간 인퓨징한다.

2. 다시 불에 올린 후 40°C 정도가 되면 젤라틴매스를 넣고 잘 섞어준다.

3. 화이트초콜릿이 담긴 볼에 체로 거른 **2**를 2~3번 나눠 부으며 섞어준다.

4. 블렌더로 유화시킨 후 냉장고에 보관한다.

라즈베리 콩포트

물	20g
설탕	25g
냉동 라즈베리	130g
NH펙틴	1g
티뭇후추	약간
올리브오일	3g

1. 냄비에 물, 설탕을 넣고 113°C까지 끓인다.
   ● 이때 여분의 설탕을 조금 남겨둔다.

2. 냉동 라즈베리를 넣고 50°C가 될 때까지 가열한다.

3. 남겨둔 설탕과 NH펙틴을 섞어 넣은 후 되직해질 때까지 몇 분간 더 끓인다.

4. 되직한 상태가 되면 불에서 내려 티뭇후추, 올리브오일을 넣고 가볍게 섞는다.

5. 지름 3.5cm 반구형 몰드에 채운 후 냉동실에서 얼려 준비한다.

Pepper Mousse

380 g	Heavy cream (36%)
2 g	Timut pepper
10 g	Gelatin mass (200bloom)
88 g	White chocolate
(Valrhona Ivoire 35%)	

1. In a saucepan, bring 1/3 of heavy cream and timut pepper to a boil, remove from the heat, and let it infuse for about 10 minutes.

2. Put the mixture back over the heat and combine gelatin mass at 40°C.

3. Add strained **2** to a bowl with white chocolate in 2~3 parts to mix.

4. Emulsify with a blender, and store in a refrigerator.

Raspberry Compote

20 g	Water
25 g	Sugar
130 g	Frozen raspberries
1 g	Pectin NH
Pinch of Timut pepper	
3 g	Olive oil

1. Add water and sugar to a saucepan, and boil to 113°C.
   ● At this point, set aside a little bit of sugar.

2. Add frozen raspberries and boil until 50°C.

3. Mix remaining sugar and pectin NH, then add to **2**. Boil a bit more until the mixture becomes thick.

4. When it becomes thick, remove from the heat; add timut pepper and olive oil and lightly mix.

5. Fill it in 3.5 cm diameter hemisphere mold, and freeze.

per Mousse

2

3

4

pberry Compote

2

4

5

화이트초콜릿 소스

화이트초콜릿   50g
(발로나 IVOIRE 35%)

카카오버터   50g

볼에 모든 재료를 넣고 중탕으로 녹인 후 블렌더로 유화시켜 체에 걸러 준비한다.

**조립 & 마무리
(후추 무스 케이크)**

미러 글레이즈(시판)   적당량

1. 냉장고에 보관한 후추 무스를 블렌더로 가볍게 휘핑한다.

2. 지름 4.5cm 반구형 몰드에 파이핑한다.

3. 미니 스패출러를 이용해 빈틈이 없도록 정리한다.

4. 얼린 라즈베리 콩포트가 정중앙에 오도록 넣고 살짝 눌러준다.

5. 다시 후추 무스를 파이핑해 몰드를 채운다.

6. 미니 스패출러로 윗면을 평평하게 정리한 후 냉동실에서 얼린다.

7. 화이트초콜릿 소스로 코팅한다.

8. 미러 글레이즈로 스프레이한 후 냉동실에 보관한다.

White Chocolate Sauce

50 g   White chocolate
(Valrhona Ivoire 35%)

50 g   Cacao butter

Place all the ingredients in a bowl and melt them over a double boiler. Emulsify with a blender, then sieve the mixture.

Build & Finish (Pepper Mousse Cake)

QS   Mirror glaze
          (ready-made)

1. Lightly whip the pepper mousse stored in the refrigerator with a blender.

2. Pipe in 4.5 cm diameter hemisphere mold.

3. Use a mini spatula to fill thoroughly so there are no gaps.

4. Place the frozen raspberry compote in the center and press lightly.

5. Pipe the pepper mousse again to fill.

6. Flatten the top evenly with a mini spatula and freeze.

7. Coat with white chocolate sauce.

8. Spray with mirror glaze and store frozen.

Build & Finish (Pepper Mousse Cake)

1

2

4

5

7

8

사블레✤

버터	65g
설탕	50g
소금	1g
노른자	25g
옥수수가루	80g
베이킹파우더	0.1g

1. 볼에 깍둑썰기한 차가운 상태의 버터, 설탕, 소금을 넣고 재료가 골고루 섞일 때까지 가볍게 믹싱한다.
2. 노른자를 조금씩 나눠 넣어가며 믹싱한다.
3. 미리 섞어둔 옥수수가루와 베이킹파우더를 넣어가며 믹싱한다.
4. 반죽 앞뒤로 투명 비닐 또는 유산지를 깐 후 1.5mm 두께로 밀어 편다.
5. 타공 매트에 올려 170℃로 예열된 오븐에서 7~8분간 굽는다.

사블레 믹스

사블레✤	105g
녹인 버터	20g
소금	약간
후추	약간

1. 식힌 사블레를 적당한 크기로 다진다.
2. 녹인 버터, 소금, 후추를 넣고 골고루 섞는다.
3. 지름 3.5cm 원형 틀에 9g씩 넣고 숟가락으로 살짝 눌러 고정시킨다.
4. 냉동실에서 얼린다.

Sablé✤

65 g	Butter
50 g	Sugar
1 g	Salt
25 g	Egg yolks
80 g	Corn powder
0.5 g	Baking powder

1. Lightly mix cold butter cut into cubes, sugar, and salt in a bowl until evenly mixed.
2. Continue to mix while adding egg yolks a little bit at a time.
3. Add previously mixed corn powder and baking powder to mix.
4. Place the dough between two sheets of plastic or parchment paper, and roll out to 1.5 mm thickness.
5. Put on the perforated mat to bake in an oven preheated to 170°C for 7~8 minutes.

Sablé Mix

105 g	Sablé✤
20 g	Melted butter
Pinch of Salt	
Pinch of Pepper	

1. Chop the cooled sablé into moderate size.
2. Mix thoroughly with melted butter, salt, and pepper.
3. Put 9 grams in a 3.5 cm diameter round ring and press lightly with a spoon to set it in place.
4. Freeze completely.

2

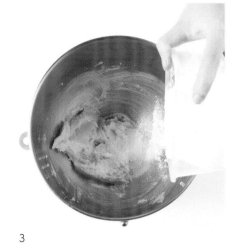

3

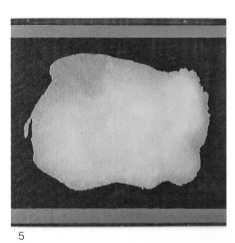

5

2

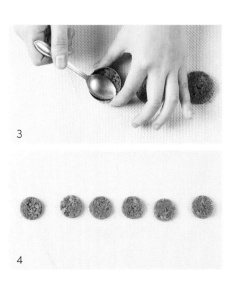

3

4

라즈베리 머랭

흰자	100g
설탕	80g
슈거파우더	75g
라즈베리파우더	30g

1. 볼에 흰자를 넣고 휘핑하다 맥주 거품 상태가 되면 설탕을 2~3번 나눠 넣으며 휘핑한다.

2. 체 친 슈거파우더 1/2을 넣고 휘핑한다.

3. 휘퍼를 들어 올렸을 때 새 부리 모양으로 크림이 올라오는 단단한 상태의 머랭이 되면 남은 슈거파우더를 2~3번 나눠 넣으며 주걱으로 조심스럽게 섞는다.

4. 투명 비닐을 깐 후 803번 깍지를 이용해 원형으로 파이핑한다.

5. 손가락으로 가볍게 터치해 긴 잎사귀 모양을 만든다.

6. 라즈베리파우더를 뿌린다.

7. 원통형 몰드 또는 구부러진 효과를 줄 수 있는 도구 위에 올려 40°C의 오븐이나 건조기에서 하루 정도 말린다.

레몬 크렘 프레슈

크렘 프레슈	50g
레몬즙	4g
레몬 제스트	적당량

1. 볼에 모든 재료를 넣고 골고루 섞는다.
   - 크렘 프레슈는 프랑스 유제품 중 하나로 유지방 함량이 45% 정도로 점도가 되직한 편이며 산미가 느껴지는 것이 특징이다. 크렘 프레슈 대신 플레인 요거트로 대체 가능하다.

2. 짤주머니에 담아 냉장고에 넣어 준비한다.

Raspberry Meringue

100 g	Egg whites
80 g	Sugar
75 g	Powdered sugar
30 g	Raspberry powder

1. Whip egg whites in a bowl until it becomes foamy, then continue to whip while adding sugar in 2~3 parts.

2. Add 1/2 of sifted powdered sugar and continue to whip.

3. When the meringue forms a stiff peak, add the rest of the powdered sugar and carefully fold with a rubber spatula in 2~3 parts.

4. Line a clear plastic sheet and pipe rounds using the No. 803 piping tip.

5. Lightly touch with a finger to make a long leaf shape.

6. Sprinkle raspberry powder.

7. Put on a cylindrical mold or a tool that can give a curved effect, and dry in an oven or dryer at 40°C for about a day.

Lemon Crème Fraîche

50 g	Crème fraîche
4 g	Lemon juice
QS	Lemon zest

1. Combine all the ingredients in a bowl and mix evenly.
   - Crème fraîche is one of the French dairy products, with a milk fat content of 45%, relatively thick, and characterized by an acidic taste. You can substitute it with plain yogurt instead of crème fraîche.

2. Put in a piping bag and refrigerate.

pberry Meringue

2

3

6

7

hon Crème Fraîche

2

플레이팅

1. 접시 한 쪽 가장자리에 라즈베리 파우더를 뿌린다.
2. 접시 정중앙에 레몬 크렘 프레슈를 소량 파이핑한 후 적당한 크기로 자른 라즈베리를 동그랗게 올려준다.
3. 사블레 믹스를 올린다.
4. 후추 무스 케이크를 올린다.
5. 후추 무스 케이크 윗면에 레몬 크렘 프레슈를 소량 파이핑한다.
6. 라즈베리 소르베를 올린다.
7. 라즈베리 소르베와 후추 무스 케이크 사이에 라즈베리 머랭을 둘러준다.
8. 라즈베리 주스는 따로 잔에 담아 테이블에서 서비스한다.

8

Plating

1. Dust raspberry powder on one side of the plate.

2. Pipe a small amount of lemon crème fraîche in the center of the plate and put raspberries cut into moderate size in a circle on top.

3. Put sablé mix over it.

4. Place pepper mousse cake on top.

5. Pipe a small amount of lemon crème fraîche on top of the pepper mousse cake.

6. Top with raspberry sorbet.

7. Encircle with meringue between raspberry sorbet and pepper mousse cake.

8. Provide raspberry juice in a separate cup and serve at the table.

Raspberry Vacherin is a dessert that I made, especially for Valentine's Day in 2021. The dessert 'Vacherin' is a French dessert made mainly of ice cream or sorbet and meringue. I wanted to make this dessert in the shape of a rose to match the Valentine's Day concept. I used raspberry to give it an intense but beautiful red color, and I paired it with timut pepper, my personal favorite spice. Timut pepper is native to Nepal and Southeast Asia, has a strong scent of flowers, and is fragrant. Also, it is a spice that goes well with berries and citrus fruits because it gives a refreshing feeling even though it's pepper. When I was in France, I used to enjoy plain yogurt, or Fromage Blanc sprinkled with a dash of vanilla sugar or honey and loaded with fresh raspberries; and to go in line with this, I served it with crème fraîche. The last point, the meringue, was made in the shape of leaves to give the impression of enveloping rose. It's a picturesque attractive dessert with a very intense red color that fits the theme of Valentine's Day.

4

/

Apple
Tatin

/

애플 타탕

타르트 타탕은 제가 가장 좋아하는 프렌치 디저트 중 하나입니다. 언젠가 저만의 방식으로 재해석해 새롭게 만들어보고 싶었고, 가을이 제철인 사과를 이용해 애플 타탕으로 만들게 되었습니다. 애플 타탕 은 사과를 큼직하게 썰어 사용하는 것이 일반적이지만 저는 좀 더 특별한 모양을 내고 싶었습니다. 그래 서 부드러우면서도 결이 주는 특별한 식감을 내기 위해 사과를 얇게 썰어 두루마리 휴지처럼 돌돌 마는 방식을 생각하게 되었습니다. 가을 사과와 어울리는 훈연 향도 내고 싶었는데요, 직접적인 향보다는 간 접적인 방법으로 향을 입히고 싶어 정산소종Lapsang Souchong을 사용해 캐러멜을 만들었습니다. (정산소 종은 훈연한 소나무 위에 두고 말리는 과정을 거치므로 훈연 향이 꽤 강하게 나는 편입니다.) 사과는 다 양한 향신료와 잘 어울리는 재료이지만 한국의 향신료를 사용하고 싶어 고민을 하며 식재료 창고를 둘 러보던 중, 한국에서 온 고운 고춧가루를 발견하고 이거다 싶어 사용하게 되었습니다. 달콤한 사과에 매 콤함이 살짝 곁들여져도 재미있겠다 싶어 고춧가루, 타르트 타탕과 찰떡궁합인 크렘 프레슈를 섞어 크 림을 만들었습니다. 본래 타르트 타탕은 푀이타주 반죽과 사과를 함께 오븐에 넣고 굽지만, 이 디저트에 서는 푀이타주 특유의 파삭함과 부서지는 식감을 살리고 싶어 반죽과 사과를 따로 구웠습니다. 푀이타 주 반죽을 메이플시럽과 설탕 위에 두고 구워 캐러멜 향을 더 진하게 느낄 수 있습니다. 스모키한 캐러 멜, 오븐에서 갓 나온 것 같은 따뜻한 상태의 애플 타탕, 차가운 코코넛 소르베는 함께 먹었을 때 잘 어우 러지며 끝맛으로 약간의 매콤함까지 느낄 수 있습니다. 프랑스를 비롯한 유럽의 손님들에게 많은 사랑 을 받았던, 쌀쌀한 가을 날씨에 먹기 좋은 warm & cold 디저트입니다.

Ingredients

10인분 ◆ SERVES 10

캐러멜 애플 타탕

설탕	160g
버터	85g
바닐라빈	1/4개
레몬 제스트	레몬 2개 분량
오렌지 제스트	오렌지 2개 분량
사과	10개 분량

Directions

1. 냄비에 설탕을 넣고 주걱으로 저어가며 캐러멜화시킨다.

2. 버터, 바닐라빈, 레몬 제스트, 오렌지 제스트를 넣고 섞는다.

3. 버터가 모두 녹으면 불에서 내려 식힌 후 체에 걸러 사용한다.

4. 껍질을 벗긴 사과를 슬라이스 기계로 얇게 저민다.

5. 지름 6cm 정도가 될 때까지 돌돌 말아준다.

    ● 구울 때 사과가 풀리지 않고 예쁘게 완성되게 하려면 최대한 빈틈 없이 타이트하게 말아주는 것이 좋다.

6. 돌돌 만 사과를 2cm 높이로 자른다.

7. 지름 6cm 원형 몰드에 **3**을 25g씩 붓는다.

8. 말아 썬 사과를 몰드에 넣은 후 몰드 위에 실팻 - 철판 순서로 올려 오븐에 넣는다. 170°C로 예열된 오븐에서 40분 굽고, 철판을 돌려 40분 더 굽는다.

    ● 철판을 올리고 구워야 모양이 흐트러지지 않는다.

Caramel Apple Tatin

160 g	Sugar
85 g	Butter
1/4	Vanilla bean
Zest of 2 Lemons	
Zest of 2 Oranges	
10	Apples

1. Add sugar to a pot and caramelize while stirring with a spatula.

2. Stir in butter, vanilla bean, lemon zest, and orange zest.

3. Once the butter is melted, remove from heat to cool and strain to use.

4. Thinly slice peeled apples with a mandoline slicer.

5. Roll the sliced apples until it is about 6 cm in diameter.

    ● To ensure the apples do not unroll during baking and are finished beautifully, it's best to roll them as tightly as possible without any gaps.

6. Cut the rolled apples to 2 cm in height.

7. Pour 25 grams of **3** into 6 cm diameter round molds.

8. Insert rolled and cut apples in the molds, then place in the order of Silpat - baking tray on top, and put in the oven. Bake in the oven preheated to 170°C for 40 minutes, rotate the baking tray, and bake for another 40 minutes.

    ● It should be baked with the tray on top to retain its shape.

2

3

5

7

8

메이플 푀이타주

중력분	280g
소금	5g
녹인 버터	45g
물	150g
바닐라파우더	0.3g
판버터	165g
메이플시럽	적당량
메이플슈거	적당량

1. 볼에 중력분, 소금, 녹인 버터, 물, 바닐라파우더를 넣고 반죽이 한 덩어리로 뭉쳐질 때까지 믹싱한다.

2. 반죽을 랩핑한 후 15분 정도 실온에서 숙성시킨다.

3. 반죽을 가로 17cm 직사각형 모양으로 밀어 편 후 판버터를 올리고 반죽을 접어 버터를 감싼다.

   ● 판버터는 17cm 정사각형 크기로 밀어 펴 사용한다.

4. 반죽을 밀어 편 후 4절 접기 1회, 3절 접기를 1회 작업한다.

5. 반죽에 비닐을 씌워 냉장고에서 1시간 정도 숙성시킨다.

6. 다시 4절 접기 1회를 작업하고 반죽을 16×40cm 크기로 밀어 편 후 1cm 폭으로 찔라 냉동실에 보관한다.

7. 실팻 위에 메이플시럽을 바른다.

8. 그 위에 메이플슈거를 뿌린다.

   ● 메이플슈거 대신 황설탕으로 대체할 수 있다.
   덩어리진 메이플슈거가 있을 수 있으므로 체에 걸러 사용하는 것이 좋다.

9. 4cm 간격으로 둔 직사각 메탈 몰드 사이에 푀이타주를 넣고 170℃로 예열된 오븐에서 23~25분 정도 굽는다.

Maple Feuilletage

280 g	All-purpose flour
5 g	Salt
45 g	Melted butter
150 g	Water
0.3 g	Vanilla powder
165 g	Butter sheet
QS	Maple syrup
QS	Maple sugar

1. Mix all-purpose flour, salt, melted butter, water, and vanilla powder in a mixing bowl until the dough comes together into a ball.

2. Plastic-wrap the dough and let it rest for about 15 minutes at room temperature.

3. Roll out the dough to a 17 cm width rectangle, place the butter sheet on top, and wrap the butter by folding the dough.

   ● Roll out the butter sheet to a 17 cm square to use.

4. Roll out the dough and make one double-fold and one single-fold.

5. Wrap the dough with plastic and rest it for about one hour in the refrigerator.

6. Make another double-fold, roll out the dough to a size of 16 × 40 cm, cut to 1 cm width, and store them in the freezer.

7. Brush maple syrup on the Silpat.

8. Sprinkle maple sugar on top.

   ● You can replace maple sugar with light brown sugar. There may be lumps of maple sugar, so it's better to sieve to use.

9. Put feuilletage between the rectangular metal molds spaced 4 cm apart, and bake in an oven preheated to 170℃ for about 23~25 minutes.

2

3

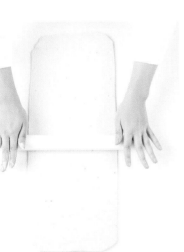

5

6

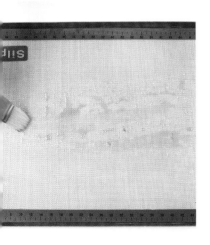

8

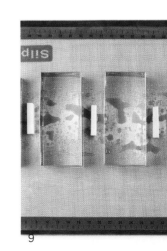

9

코코넛 소르베

물	45g
코코넛밀크	100g
설탕	55g
아이스크림 안정제 (stabilizer)	1g
코코넛 퓌레	450g

1. 냄비에 물, 코코넛밀크를 넣고 끓어오를 때까지 가열한다.
2. 설탕, 아이스크림 안정제를 넣고 끓인다.
3. 불에서 내려 코코넛 퓌레가 담긴 볼에 담는다.
4. 블렌더로 믹싱해 냉장고에서 하루 동안 숙성시킨 후 아이스크림 머신에서 소르베로 만든다.
   ● 아이스크림 머신 대신 파코젯Pacojet을 사용해도 좋다.

정산소종 캐러멜

생크림(36%)	100g
우유	40g
글루코스A	45g
소금	0.5g
정산소종 찻잎	1g
설탕	40g
글루코스B	10g
버터	30g

1. 냄비에 생크림, 우유, 글루코스A, 소금, 정산소종 찻잎을 넣고 끓어오를 때까지 가열한 후 불에서 내려 10분간 인퓨징한다.
2. 다른 냄비에 설탕, 글루코스B를 넣고 가열해 캐러멜화시킨다.
3. 1을 조금씩 넣어가며 농도가 걸쭉해질 때까지 가열한다.
4. 체에 거른다.
5. 버터를 넣고 블렌더로 유화시켜 냉장고에 보관한다.

Coconut Sorbet

45 g	Water
100g	Coconut milk
55 g	Sugar
1 g	Ice cream stabilizer
450 g	Coconut puree

1. Heat water and coconut milk in a saucepan until it boils.
2. Add sugar and ice cream stabilizer, and bring to boil.
3. Remove from the heat and pour into a bowl with coconut puree.
4. Mix with a blender, refrigerate for a day and make sorbet with the ice cream maker.
   ● If you don't have an ice cream maker, you can also use Pacojet.

Lapsang Souchong Caramel

100 g	Heavy cream (36%)
40 g	Milk
45 g	Glucose A
0.5 g	Salt
1 g	Lapsang souchong tea leaves
40 g	Sugar
10 g	Glucose B
30 g	Butter

1. Heat heavy cream, milk, glucose A, salt, and lapsang souchong tea leaves in a saucepan until it boils, then remove from the heat and let it infuse for 10 minutes.
2. In a different pot, heat sugar and glucose B to caramelize.
3. Add 1 little by little and heat until thickened.
4. Filter through a sieve.
5. Add butter, emulsify using a blender, then store in the refrigerator.

onut Sorbet

2

3

4

sang Souchong Caramel

2

3

4

5

캐러멜 시럽

설탕	50g
물	60g
바닐라빈	1/4개
오렌지 제스트	오렌지 1개 분량
레몬 제스트	레몬 1개 분량
버터	10g

1. 냄비에 설탕을 넣고 175°C까지 가열하여 캐러멜화시킨다.
2. 물, 바닐라빈, 오렌지 제스트, 레몬 제스트를 넣고 가열한다.
3. 버터를 넣고 녹을 때까지 가열한 후 체에 걸러 식혀 냉장고에 보관한다.

스파이시 크림

크렘 프레슈	100g
고운 고춧가루	0.5g
소금	1q
레몬즙	7g
레몬 제스트	적당량
파넬라슈거	3g

1. 볼에 모든 재료를 넣는다.
   - 크렘 프레슈는 프랑스 유제품 중 하나로 유지방 함량이 45% 정도로 점도가 되직한 편이며 산미가 느껴지는 것이 특징이다. 크렘 프레슈 대신 플레인 요거트로 대체 가능하다.
   - 고운 고춧가루는 정확한 계량이 필요하므로 미량 저울을 이용해 계량한다.
   - 파넬라슈거는 멕시코의 비정제 설탕으로 블록 형태라 강판에 갈아 사용해야 한다. 특유의 캐러멜 향이 진해 개인적으로 좋아하는 재료 중 하나이며 황설탕으로 대체해도 좋다.
2. 휘퍼로 골고루 섞는다.
   - 크렘 프레슈는 유지방 함량이 높아 너무 많이 섞게 되면 분리가 일어날 수 있으니 주의한다.
3. 짤주머니에 담아 냉장고에 보관한다.

Caramel Syrup

50 g	Sugar
60 g	Water
1/4	Vanilla bean
	Zest of 1 Orange
	Zest of 1 Lemon
10 g	Butter

1. Put sugar in a pot and caramelize it to 175°C.
2. Add water, vanilla bean, orange zest, and lemon zest, and heat.
3. Add butter and heat until it melts, strain to cool, and store in the refrigerator.

Spicy Cream

100 g	Crème fraîche
0.5 g	Fine red pepper powder
1 g	Salt
7 g	Lemon juice
QS	Lemon zest
3 g	Panela sugar

1. Put all the ingredients in a bowl.
   - Crème fraîche is one of the French dairy products, with a milk fat content of 45%, relatively thick, and characterized by an acidic taste. You can substitute it with plain yogurt instead of crème fraîche.
   - Since fine red pepper powder requires precise measurement, use a micro-scale.
   - Panela sugar is unrefined sugar from Mexico in a block form, so it has to be ground on a grater to use. It is one of the ingredients often used for its unique flavor and caramel aroma. If it's unavailable, you can replace it with brown sugar.
2. Combine thoroughly with a whisk.
   - Because crème fraîche has high milk fat content if you stir too much cream may separate.
3. Pour in a piping bag and refrigerate.

cy Cream

2

3

플레이팅

1. 접시 오른쪽 부분에 정산소종 캐러멜을 두 군데 파이핑한다.

2. 정산소종 캐러멜 사이에 스파이시 크림을 파이핑한다.

3. 스파이시 크림 위에 메이플 퓌이타주를 올린다.

4. 오븐에서 따뜻한 상태로 데운 캐러멜 애플 타탕에 캐러멜 시럽을 발라 윤기가 나게 한 후 메이플 퓌이타주 옆에 올린다.

5. 메이플 퓌이타주 위에 코코넛 소르베를 올린다.

Plating

1. Pipe two Lapsang Souchong caramel rounds on the right side of the plate.

2. Pipe spicy cream between the caramel rounds.

3. Put maple feuilletage over the spicy cream.

4. Brush caramel syrup on the caramel apple tatin warmed in the oven to make it glossy, and then place it next to the maple feuilletage.

5. Top with coconut sorbet over the maple feuilletage.

Tart Tatin is one of my favorite French desserts. I wanted to reinterpret it in my own way one day to make it new, and I did get to make the apple tatin using apples that are in season in fall. It's common to cut apples into big pieces for apple tatin, but I wanted to give it a more special shape. So, I came up with the idea of slicing apples thinly and rolling them up like a roll of tissue in order to create something soft and yet have a special texture by stacking them in many thin layers. I wanted to create a smoky flavor that goes well with the autumn apples, but I wanted to give it an indirect scent rather than directly, so I made caramel using Lapsang Souchong. (Lapsang Souchong tea leaves are placed on a smoked pine tree to dry; therefore, the smoky aroma is quite strong.) Apple is an ingredient that goes well with various spices, but while I was looking around the food ingredient warehouse wanting to use Korean spice, I found the fine red pepper powder from Korea and thought this was it. I thought adding a little bit of spiciness to the sweet apple would be fun, so I mixed the red pepper powder with crème fraîche, which is a perfect pair with tart tatin. Originally, feuilletage dough and apples are baked together in an oven to make tart tatin. But, for this dessert, I baked the dough and apples separately to preserve the crisp and flaky texture of the feuilletage. Feuilletage dough is baked over maple syrup and sugar, so that you can enjoy the caramel flavor even more. When eaten together, the smoky caramel, warm apple tatin as if it's just out of the oven, and cold coconut sorbet goes well together, and you can feel a hint of spiciness at the end. It is a warm & cold dessert that is delightful to have in the chilly autumn weather, which French and European customers adored.

5
/
Baby Banana
/
베이비 바나나

정식당의 인기 메뉴이자 저의 대표 시그니처 메뉴이기도 한 '베이비 바나나'입니다. 이 디저트는 2017년부터 만들기 시작했는데 당시 한국에서 유행하던 다양한 종류의 바나나 스낵들과 뉴요커들이 사랑하는 바나나 푸딩에서 영감을 얻었습니다. (저 역시 바나나를 사용한 스낵이나 디저트를 굉장히 좋아합니다.) 처음에는 바나나 푸딩으로 포스트 디저트 서비스를 했는데 손님들의 반응이 너무 좋아 바나나를 사용한 메인 디저트를 개발하게 되었고, 특별함을 더하기 위해 진짜 바나나와 똑같은 모양으로 만들었습니다. (프랑스의 유명 셰프인 세드릭 그롤레 옆에서 오랜 시간 동안 함께 일하며 배운 테크닉을 접목시켜본 메뉴이기도 합니다.) 함께 곁들인 아이스크림은 바나나 푸딩으로 포스트 디저트 서비스를 하던 때에 많은 손님들이 바나나 푸딩과 커피를 함께 드시는 것을 떠올려 커피 아이스크림으로 만들었고, 진짜 바나나 같은 귀여운 모양의 디저트에 좀 더 특별한 재미를 주기 위해 신선한 제철과일을 가득 담은 바구니 위에 베이비 바나나를 올려 손님들이 과일 바구니에서 베이비 바나나를 직접 집어 본인의 접시에 담을 수 있도록 서비스했습니다.

식용 실리콘 몰드를 이용해 베이비 바나나 틀을 직접 제작해 몇 년을 사용해오다가 2020년 이탈리아 몰드 회사의 제안을 받아 제 이름으로 베이비 바나나 몰드를 출시해 몰딩하는 것이 훨씬 수월해졌습니다. 베이비 바나나 디저트로 2019년 'Rising star'와 'Art of Presentation' 상도 받게 되었고, 다양한 행사에도 많이 참여해 제 이름을 뉴욕에 더 많이 알릴 수 있게 되어 저에게 아주 큰 의미가 있는 디저트입니다.

Ingredients

커피 아이스크림

우유	250g
생크림(36%)	125g
원두가루	8g
황설탕	60g
노른자	100g
버터	33g

Coffee Ice Cream

250 g	Milk
125 g	Heavy cream (36%)
8 g	Ground coffee beans
60 g	Light brown sugar
100 g	Egg yolks
33 g	Butter

Directions

1. 냄비에 우유, 생크림, 원두가루를 넣고 랩을 씌워 하루 동안 냉장 보관하며 인퓨징한다.
2. 중불에서 휘퍼로 저어가며 끓인다.
3. 황설탕을 넣고 끓인다.
4. 노른자를 넣고 휘퍼로 저어가며 끓여 크렘 앙글레이즈를 만든다.
5. 체에 거른다.
6. 버터를 넣고 블렌더로 유화시킨 후 냉장고에서 하루 동안 숙성시킨다. 사용하기 전 블렌더로 한 번 더 믹싱한 후 아이스크림 머신에 넣고 돌린다.

   ● 아이스크림 머신 대신 파코젯Pacojet을 사용해도 좋다.

1. Put milk, heavy cream, and ground coffee beans in a pot, cover it with plastic wrap, and leave it to infuse for a day in a refrigerator.
2. Boil over medium heat while stirring with a whisk.
3. Add light brown sugar and boil.
4. Add egg yolks and boil while stirring with a whisk to make crème Anglaise.
5. Filter through a sieve.
6. Add butter and emulsify with a blender, then store in a refrigerator for a day. Mix with a blender once more before use and run it in an ice cream maker.

   ● If you don't have an ice cream maker, you can also use Pacojet.

둘세 가나슈

생크림(36%)	390g
바닐라빈	1/4개
젤라틴매스 (200bloom)	10g
블론드 초콜릿 VALRHONA BLOND DULCEY(32%)	85g

1. 냄비에 생크림 1/3과 바닐라빈을 넣고 끓어오를 때까지 가열한 후 불에서 내려 볼 입구에 랩을 씌워 10분 정도 인퓨징한다.
2. 젤라틴매스를 넣고 중불에서 녹을 때까지 가열한다.
3. 블론드 초콜릿에 2를 두 번에 나눠 부어가며 저어준다.
4. 남은 생크림을 붓고 휘퍼로 가볍게 저어준다.
5. 체에 거른다.
6. 블렌더로 유화시킨 후 밀착 랩핑해 하루 동안 냉장고에 보관한다.
   - 완성된 둘세 가나슈는 냉장고에서 3일간 보관하며 사용할 수 있다.

바나나 베일리스 밀크

우유	100g
바나나	50g
황설탕	7g
꿀	5g
베일리스(오리지널) BAILEYS (ORIGINAL)	18g

1. 냄비에 우유, 적당한 크기로 자른 바나나를 넣고 끓어오를 때까지 가열한다.
2. 불에서 내려 황설탕, 꿀, 베일리스를 넣고 블렌더로 믹싱한 후 밀착 랩핑해 냉장고에 보관한다.
   - 완성된 바나나 베일리스 우유는 냉장고에서 2일간 보관하며 사용할 수 있다.

Dulcey Ganache

390 g	Heavy cream (36%)
1/4	Vanilla bean
10 g	Gelatin mass (200bloom)
85 g	Blonde chocolate (Valrhona Blonde Dulcey 32%)

1. Heat 1/3 of heavy cream and vanilla bean in a saucepan until it boils, then cover it with plastic wrap and infuse for about 10 minutes.
2. Add gelatin mass and heat over medium heat until it dissolves.
3. Pour 2 over the Blonde chocolate in two portions while stirring.
4. Pour the remaining heavy cream and stir lightly.
5. Filter through a sieve.
6. Emulsify with a blender, then wrap the cream, making sure the wrap adheres to the cream, and refrigerate for one day.
   - The completed Dulcey Ganache can be stored in the refrigerator and used for 3 days.

Banana Baileys Milk

100g	Milk
50 g	Banana
7 g	Light brown sugar
5 g	Honey
18 g	Baileys (Original)

1. Heat milk and bananas cut into moderate size in a pot to a boil.
2. Remove from the heat, and mix light brown sugar, honey, and Baileys with a blender. Wrap it, making sure the wrap adheres to the mixture, and refrigerate.
   - The completed Banana Baileys Milk can be stored in the refrigerator and used for 2 days.

Icey Ganache

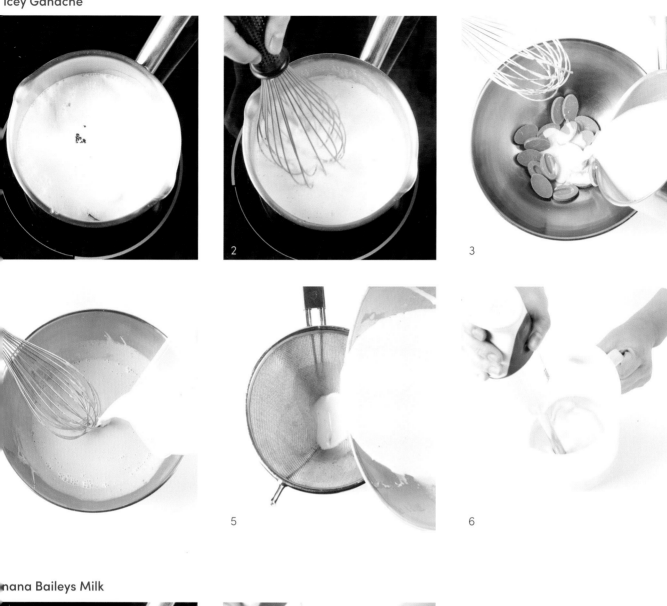

nana Baileys Milk

바나나 크레뮤

생크림(36%)	43g
우유	43g
바나나	210g
설탕	13g
황설탕	10g
노른사	48g
셀라틴매스 (200bloom)	8g

1. 냄비에 생크림, 우유, 적당한 크기로 자른 바나나를 넣고 블렌더로 갈아준다.
   ● 바나나는 잘 익은 것을 사용한다.
2. 볼에 설탕, 황설탕, 노른자를 넣고 휘퍼로 골고루 섞는다.
3. 2에 1을 넣고 섞는다.
4. 볼에 올려 휘피로 지이기머 83~84℃까지 끓인다.
5. 물에서 내려 젤라틴매스를 넣고 섞는다.
6. 블렌더로 믹싱한 후 밀착 랩핑해 냉장고에 보관한다.
   ● 완성된 바나나 크레뮤는 냉장고에서 3일간 보관하며 사용할 수 있다.

Banana Cremeux

43 g	Heavy cream (36%)
43 g	Milk
210 g	Banana
13 g	Sugar
10 g	Light brown sugar
48 g	Egg yolks
8 g	Gelatin mass (200bloom)

1. Put heavy cream, milk, and bananas cut into moderate size in a pot and grind with a blender.
   ● Use ripe bananas.
2. Stir sugar, light brown sugar, and egg yolks thoroughly in a bowl with a whisk.
3. Add 1 into 2 and mix.
4. Boil them until 83~84°C while stirring with a whisk.
5. Remove from the heat and stir in gelatin mass.
6. Mix with a blender, then wrap it, making sure the wrap adheres to the cremeux, and refrigerate.
   ● The completed Banana Cremeux can be stored in the refrigerator and used for 3 days.

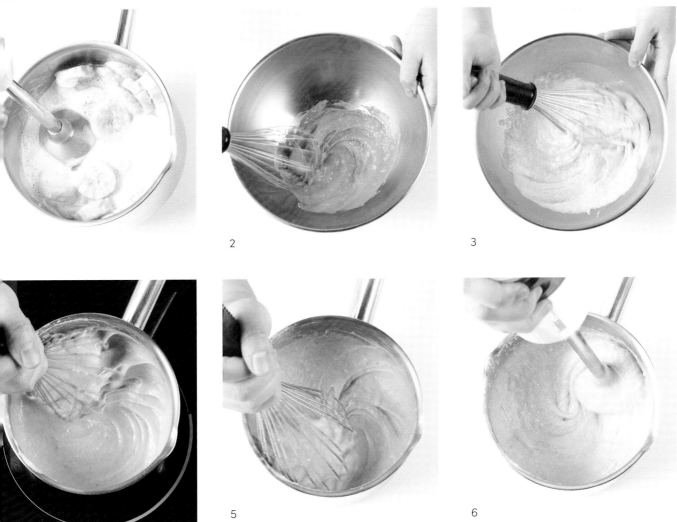

2

3

5

6

바나나 케이크

달걀	4개
설탕	170g
생크림(36%)	90g
베일리스(오리지널)	70g
BAILEYS (ORIGINAL)	
바나나	150g
브라운 버터	70g
박력분	150g
베이킹파우더	4g

1. 볼에 달걀, 설탕을 넣고 골고루 섞는다.

2. 다른 볼에 생크림, 베일리스, 적당한 크기로 자른 바나나를 넣고 블렌더로 믹싱한다.

3. 1과 2를 섞는다.

4. 브라운 버터를 넣고 섞는다.

   ● 브라운 버터는 냄비에서 약불로 갈색이 될 때까지 가열해 사용한다.

5. 체 친 박력분, 베이킹파우더를 넣고 섞는다.

6. 유산지를 깐 팬에 붓고 170°C로 예열된 오븐에서 50분간 굽는다.

   ● 뾰족한 도구로 반죽 가운데를 찔렀을 때 반죽이 묻어나오지 않으면 잘 익은 것이다.

7. 구워져 나온 케이크는 팬에서 꺼내 식힌 후 가로 2cm, 세로 7cm, 두께 0.5cm로 자른다.

   ● 완성된 바나나 베일리스 케이크는 2주 동안 냉동실에서 보관하며 사용할 수 있다.

Banana Cake

4	Eggs
170 g	Sugar
90 g	Heavy cream (36%)
70 g	Baileys (Original)
150 g	Banana
70 g	Brown butter
150 g	Cake flour
4 g	Baking powder

1. Mix eggs and sugar thoroughly in a bowl.

2. In a separate bowl, mix heavy cream, Baileys, and bananas cut into moderate size with a blender.

3. Combine 1 and 2.

4. Mix with brown butter.

   ● Cook butter while stirring until it turns light brown.

5. Mix with sifted cake flour and baking powder.

6. Pour the batter onto a baking pan lined with parchment paper. Bake in an oven preheated to 170°C for 50 minutes.

   ● When you stick a pointy tool in the center of the cake, it is cooked if it comes out clean.

7. Remove the cake from the pan to cool, and cut it into 2 cm wide, 7 cm long, and 0.5 cm thick.

   ● The completed Banana Baileys Cake can be stored in the freezer and used for 2 weeks.

2

3

5

6

몽타주

1. 냉장고에 보관한 둘세 가나슈를 가볍게 휘핑한다.
2. 바나나 모양 몰드에 파이핑한다.
   - 여기에서는 Pavoni사의 Baby Banana 몰드를 사용했다.
3. 미니 스패출러로 둘세 가나슈가 바나나 몰드 가장자리까지 빈틈 없이 채워지도록 정리한다.
   - 바나나 케이크가 들어갈 가운데 부분은 옴폭하게 만든다.
4. 바나나 크레뮤를 파이핑한다.
5. 바나나 베일리스 밀크에 적신 바나나 케이크를 넣고 살짝 눌러준다.
6. 둘세 가나슈를 바나나 크기 절반 정도로 파이핑한다.
7. 미니 스패출러로 평평하게 정리한다.
8. 냉동실에서 얼린다.

Montage

1. Lightly whip the Dulcey ganache stored in the refrigerator.
2. Pipe into the banana-shaped mold.
   - Here, I used the Baby Banana mold by Pavoni.
3. Organize using a mini spatula to make sure Dulcey ganache is filled to the edge of the banana mold without any gaps.
   - Make the middle part dented where the banana cake will go.
4. Pipe banana cremeux.
5. Insert banana cake soaked in Baileys milk and press lightly.
6. Pipe Dulcey ganache about half the size of the banana mold.
7. Spread evenly with a mini spatula.
8. Freeze completely.

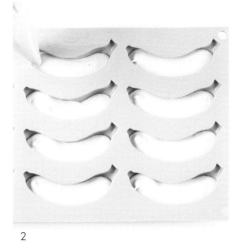

2

3

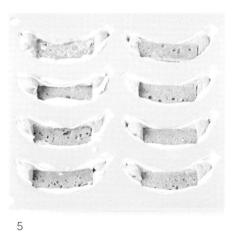

5

6

8

초콜릿 헤이즐넛 크럼블

파에테포요틴	110g
건조 바나나 칩	50g
헤이즐넛	90g
녹인 밀크초콜릿 VALRHONA JIVARA(40%)	110g
헤이즐넛아몬드프랄린 페이스트	100g
게랑드 소금	3g

1. 볼에 적당한 크기로 다진 파에테포요틴과 건조 바나나칩, 헤이즐넛,
녹인 밀크 초콜릿을 준비한다.

   ● 헤이즐넛과 건조 바나나칩은 파에데포요틴과 비슷한 크기로 다져 준비한다.
   헤이즐넛은 170℃로 예열된 오븐에서 10분 정도 구운 후 식혀 사용하면 향이 훨씬 더 좋다.

2. 헤이즐넛아몬드프랄린 페이스트, 게랑드 소금을 넣고 골고루 섞는다.

노란색 디핑 초콜릿

화이트초콜릿 VALRHONA IVOIRE(35%)	100g
카카오버터	100g
노란색 지용성 색소	0.8g

볼에 모든 재료를 넣고 중탕으로 녹인 후 블렌더로 유화시킨다.

● 완성된 디핑 초콜릿은 실온에 보관하며, 3주 동안 사용할 수 있다.

Chocolate Hazelnut Crumble

110 g	Paillete feuilletine
50 g	Dried banana chips
90 g	Hazelnuts
110 g	Milk chocolate, melted (Valrhona Jivara 40%)
100 g	Hazelnut almond praline paste
3 g	Guérande salt

1. Prepare chopped paillete feuilletine, dried banana chips and
hazelnuts to moderate size, and melted milk chocolate in a bowl.

   ● Chop hazelnuts and dried banana chips to a size similar to paillete feuilletine.
   Hazelnuts have a much better flavor when baked in an oven preheated to 170℃
   for 10 minutes and cooled to use.

2. Add hazelnut almond praline paste and Guérande salt and mix
thoroughly.

Yellow Dipping Chocolate

100 g	White chocolate (Valrhona Ivoire 35%)
100 g	Cacao butter
0.8 g	Yellow chocolate coloring

Put all ingredients in a bowl and melt over a double boiler. Mix with a
blender to emulsify.

● Completed Dipping Chocolate can be stored at room temperature and used for
3 weeks.

colate Hazelnut Crumble

2

ow Dipping Chocolate

밀크 디핑 초콜릿

밀크초콜릿	50g
VALRHONA JIVARA(40%)	
카카오버터	50g

볼에 모든 재료를 넣고 중탕으로 녹인 후 블렌더로 믹싱해 사용한다.

● 완성된 디핑 초콜릿은 실온에서 1개월간 보관하며 사용할 수 있다.

마무리

코코아파우더	적당량

1. 얼린 베이비 바나나는 몰드에서 빼낸 후 나이프로 매끈하게 정리한다.

2. 베이비 바나나에 꼬치를 꽂은 후 노란색 디핑 초콜릿에 디핑한다.

   ● 얇게 코팅될 수 있도록 디핑한 후 가볍게 털어준다.

3. 비닐을 깐 철판에 올린 후 냉장고에 보관한다.

4. 밀크 디핑 초콜릿을 스프레이 건에 넣고 분사한다.

   ● 바나나 꼭지와 아랫부분에 조금 더 분사해 음영을 준다.

5. 코코아파우더를 묻혀 좀 더 사실감 있게 마무리한다.

6. 완성된 베이비 바나나는 냉장고에 보관한다.

Milk Dipping Chocolate

50 g	Milk chocolate
(Valrhona Jivara 40%)	
50 g	Cacao butter

Put all ingredients in a bowl and melt over a double boiler, and mix with a blender to use.

● Completed Dipping Chocolate can be stored at room temperature and used for 1 month.

Finish

QS	Cocoa powder

1. Remove the frozen Baby Bananas from the mold and trim the edge smooth with a paring knife.

2. Stick a skewer in a Baby Banana and dip it in yellow dipping chocolate.

   ● Shake lightly after dipping so that it can be coated thinly.

3. Place on a baking tray lined with a plastic sheet and freeze.

4. Put milk dipping chocolate in the spray gun and spray.

   ● Spray a little more on the tip and bottom of the banana to give a shading effect.

5. Brush cocoa powder for a more realistic finish.

6. Store the finished Baby Bananas in a freezer.

2

3

5

6

플레이팅	접시에 초콜릿 헤이즐넛 크럼블을 올린 후 그 위에 커피 아이스크림 한 스쿱을 올린다. 그 옆에 베이비 바나나를 놓는다.
Plating	Place chocolate hazelnut crumble on a plate and put one scoop of coffee ice cream on top. Arrange a Baby Banana next to it.

'Baby Banana' is a popular menu at Jungsik and also my signature menu. It was in 2017 when I started making this dessert, and I was inspired by various kinds of banana snacks that were popular in Korea at that time and banana pudding New Yorkers loved in the States. (I also adore snacks and desserts made with bananas.) At first, I did a post-dessert service with banana pudding, but the response from customers was so good that we developed the main dessert using bananas and shaped it like a real banana to make it special. (It's also a menu that applies the techniques learned from the famous French chef Cedric Grolet, with whom I worked together for a long time.) As for the ice cream served with banana pudding, I recalled a lot of customers having banana pudding and coffee together during post-dessert service, so I decided to make coffee ice cream. To give an extraordinary pun to this cute dessert that looks like a real banana, I placed Baby Banana on a basket full of fresh seasonal fruits and serviced to let the customers pick the dessert themselves onto their own plates.

I made my own Baby Banana mold using a food-grade silicone mold and used it for several years, and in 2020, I released the Baby Banana mold under my name at the suggestion of the Italian mold company. This made the work much more manageable. With the Baby Banana dessert, I received the 2019 'Rising star' and 'Art of Presentation', and it is a dessert that has a great meaning to me as it has been an opportunity to publicize my name to the New York scene by participating in various events.

6

/

Lotus
Flower

/

연꽃

시트러스류의 과일들을 사랑하는 저에게 겨울은 항상 설레고 기다려지는 계절입니다. 겨울철 시트러스류의 과일을 사용한 디저트를 만들고 싶어 고민하던 중, 새콤 달콤 상큼한 과일들과 최고의 궁합을 자랑하는 머랭으로 만든 바슈랭을 조합해 보면 어떨까 생각하게 되었습니다. '바슈랭'이라는 디저트는 이미 많은 셰프들에 의해 다양한 방식으로 만들어져 왔기 때문에 어떻게 하면 나만의 방식으로 새롭게 만들 수 있을지에 대한 고민을 많이 했습니다. 그러던 중, 프랑스 시절부터 '이런 것들은 나중에 디저트로 꼭 표현해봐야지.'라고 생각하며 적어두었던 아이디어 노트를 보게 되었고, 거기에 있는 연등과 연꽃에 관한 아이디어에서 모티브를 얻어 머랭을 파이핑하는 방식에 대한 고민과 여러 시도 끝에 연꽃이라는 디저트를 완성할 수 있었습니다. 이 디저트 또한 시트러스류의 과일이 주인공이기 때문에 최대한 다양한 종류를 사용하려고 했습니다. 이 디저트에서 사용된 크림, 소르베, 잼, 콩피, 세그먼트 등에 시트러스를 이용한 다양한 조리 방식을 담아보았습니다. 연꽃 모양으로 파이핑한 얇은 머랭을 덮어 마무리해 숟가락으로 탁 쳐 깨트려 먹는 재미도 느낄 수 있는 디저트입니다.

Ingredients

10인분 ◆ SERVES 10

연꽃 머랭

흰자	100g
설탕	80g
슈거파우더	65g
오일	적당량

Directions

1. 볼에 흰자를 넣고 맥주 거품 상태가 될 때까지 가볍게 휘핑한다.
2. 설탕을 조금씩 넣어가며 휘핑한다.
3. 휘퍼를 들어 올렸을 때 새 부리 모양으로 크림이 올라오는 단단한 상태의 머랭이 될 때까지 휘핑한다.
4. 슈거파우더를 두세 번 나눠 넣어가며 섞는다.
5. 오일 스프레이를 얇게 뿌려준 지름 5cm 반구형 몰드에 연꽃 모양으로 머랭을 파이핑한다.
   - 오일 스프레이 대신 식용유를 얇게 발라도 좋다.
6. 40°C로 예열된 오븐에서 말리듯 4시간 동안 굽는다.
   - 완성된 연꽃 머랭은 몰드에서 분리한 후 건조기 또는 같은 온도의 오븐에 보관한다.

Lotus Flower Meringue

100 g	Egg whites
80 g	Sugar
65 g	Powdered sugar
QS	Oil

1. Lightly whip egg whites in a bowl until it becomes foamy.
2. Continue to whip while adding sugar little by little.
3. Whip to make a firm meringue until it forms a stiff peak.
4. Mix powdered sugar while adding in two to three times.
5. Pipe meringues in the shape of the lotus flowers on 5 cm hemisphere molds sprayed lightly with oil.
   - You can brush a thin layer of vegetable oil instead of spraying with oil.
6. Bake in an oven preheated to 40°C to dry for 4 hours.
   - Separate the finished meringues from the mold, and store them in an oven at the same temperature or in a dehydrator.

사블레 브르통

버터	75g
설탕	65g
소금	1g
노른자	30g
중력분	100g
베이킹파우더	2g

1. 볼에 차가운 상태의 깍둑 썬 버터, 설탕, 소금을 넣고 재료가 골고루 섞일 때까지 믹싱한다.
2. 노른자를 두 번에 나눠 넣어가며 골고루 섞일 때까지 믹싱한다.
3. 체 친 중력분, 베이킹파우더를 넣고 반죽이 한 덩어리가 될 때까지 믹싱한다.
4. 반죽을 두께 0.5cm로 밀어 편 후 냉장고에 잠시 둔다.
5. 반죽을 지름 6cm 원형 틀로 자른다.
6. 반죽 정중앙을 지름 3cm 원형 틀로 자른다.
7. 타공 매트 위에 올린 후 170°C로 예열된 오븐에서 10~12분간 굽는다.

Sablé Breton

75 g	Butter
65 g	Sugar
1 g	Salt
30 g	Egg yolks
100 g	All-purpose flour
2 g	Baking powder

1. Mix cold butter cut into cubes, sugar, and salt in a mixing bowl until evenly mixed.
2. Add egg yolks in two portions while mixing. Mix evenly after each addition.
3. Mix with sifted all-purpose flour and baking powder until the dough forms a ball.
4. Roll out to 0.5 cm and refrigerate for a while.
5. Cut the dough with a 6 cm round cutter.
6. Make a hole in the center with a 3 cm round cutter.
7. Put them on the perforated mat, and bake in an oven preheated to 170°C for about 10~12 minutes.

2

4

6

7

칼라만시 크림

칼라만시 원액	90g
꿀	2g
설탕	100g
달걀	105g
젤라틴매스(200bloom)	13g
버터	110g

1. 냄비에 칼라만시 원액, 꿀을 넣고 끓인다.
2. 미리 섞어둔 설탕과 달걀이 담긴 볼에 넣고 골고루 섞는다.
3. 다시 불에 올려 크림 상태로 걸쭉해질 때까지 끓인다.
4. 체에 거른다.
5. 녹인 젤라틴매스, 버터를 넣고 블렌더로 믹싱한 후 냉장고에 보관한다.

오렌지 콩피

카라카라오렌지	1개
물	250g
설탕	150g
글루코스(시럽)	20g

1. 냄비에 차가운 물(분량 외), 오렌지 껍질을 넣고 끓어오르면 체에 거른다. 이 과정을 총 3번 반복한다.
2. 냄비에 물, 설탕, 글루코스를 넣고 끓인다.
3. 1에 2를 붓고 유산지로 덮어준 후 3~4일간 실온에 두고 설탕(분량 외)을 조금씩 추가하면서 70°C 온도로 데워주는 것을 반복한다.
   - 오렌지 콩피가 투명해지고 부드러워질 때까지 반복한 후 냉장고에 보관한다.

Kalamansi Cream

90 g	Kalamansi concentrate
2 g	Honey
100 g	Sugar
105 g	Eggs
13 g	Gelatin mass (200bloom)
110 g	Butter

1. Boil kalamansi concentrate and honey in a saucepan.
2. Pour into the bowl with previously mixed sugar and eggs, and mix thoroughly.
3. Bring back over the heat and boil until it thickens to a creamy state.
4. Filter through a sieve.
5. Mix melted gelatin mass and butter with a blender and store in a refrigerator.

Orange Confit

1	Cara Cara orange
250 g	Water
150 g	Sugar
20 g	Glucose (syrup)

1. Put cold water (other than the requested) and orange peel in a pot, and when it boils, strain through a sieve. Repeat this process a total of three times.
2. Boil water, sugar, and glucose in a saucepan.
3. Pour 2 into 1 and cover with a parchment paper, and while leaving at room temperature for 3~4 days, repeat adding sugar a little bit at a time, and heat to 70°C.
   - Repeat until the orange confit is translucent and soft, then store in the refrigerator.

2

3

5

nge Confit

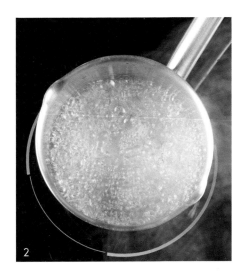

2

3

레몬 콩피

마이어레몬	100g
물	48g
칼라만시 원액	36g
글루코스(시럽)	6g

1. 냄비에 차가운 물(분량 외), 레몬 껍질을 넣고 끓어오르면 체에 거른다. 이 과정을 총 5번 반복한다.

2. 냄비에 물, 칼라만시 원액, 글루코스를 넣고 끓인다.

3. 1에 2를 붓고 유산지로 덮어준 후 3~4일간 실온에 두고 설탕을 조금씩 추가하면서 70°C 온도로 데워주는 것을 반복한다.
   ● 레몬 콩피가 투명해지고 부드러워질 때까지 반복한 후 냉장고에 보관한다.

금귤 콩피

금귤	40g
물	100g
설탕	60g
글루코스(시럽)	8g

1. 냄비에 차가운 물(분량 외), 반으로 자른 금귤을 넣고 끓어오르면 체에 거른다. 이 과정을 총 3번 반복한다.

2. 냄비에 물, 설탕, 글루코스를 넣고 끓인다.

3. 1에 2를 붓고 유산지로 덮어준 후 3~4일간 실온에 두고 설탕을 조금씩 추가하면서 70°C 온도로 데워주는 것을 반복한다.
   ● 금귤 콩피가 투명해지고 부드러워질 때까지 반복한 후 냉장고에 보관한다.

Lemon Confit

100 g	Meyer lemon
48 g	Water
36 g	Kalamansi concentrate
6 g	Glucose (syrup)

1. Put cold water (other than the requested) and lemon peel in a pot, and when it boils, strain through a sieve. Repeat this process a total of five times.

2. Boil water, kalamansi concentrate, and glucose in a saucepan.

3. Pour 2 into 1 and cover with a parchment paper, and while leaving at room temperature for 3~4 days, repeat adding sugar a little bit at a time, and heat to 70°C.
   ● Repeat until the lemon confit is translucent and soft, then store in the refrigerator.

Kumquat Confit

40 g	Kumquats
100 g	Water
60 g	Sugar
8 g	Glucose (syrup)

1. Put cold water (other than the requested) and halved kumquats in a pot, and when it boils, strain through a sieve. Repeat this process a total of three times.

2. Boil water, sugar, and glucose in a saucepan.

3. Pour 2 into 1 and cover with a parchment paper, and while leaving at room temperature for 3~4 days, repeat adding sugar a little bit at a time, and heat to 70°C.
   ● Repeat until the kumquat confit is translucent and soft, then store in the refrigerator.

on Confit

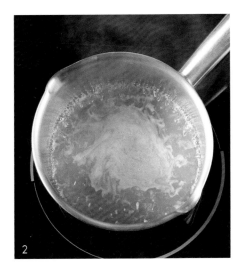

2

3

nquat Confit

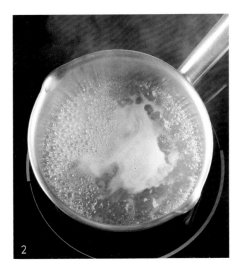

2

3

만다린 콩포트

만다린	280g
만다린즙	30g
설탕	90g
펙틴	3g
물	6g
전분	3g

1. 만다린을 껍질과 과육으로 분리한다.
   - 여기에서는 한국에서 구하기 쉬운 한라봉을 사용했다. 취향에 따라 오렌지, 천혜향 등으로 대체해도 좋다.
2. 냄비에 차가운 물(분량 외), 만다린 껍질을 넣고 끓어오르면 체에 거른다. 이 과정을 총 3번 반복한 후 블렌더로 곱게 간다.
3. 냄비에 만다린 과육과 즙, 2를 넣고 끓인 후 블렌더로 믹싱해 냉장고에 보관한다.
   - 내용물이 너무 잘게 갈리지 않도록 주의한다.
4. 미리 섞어둔 설탕과 펙틴을 넣고 끓인다.
   - 설탕과 펙틴을 섞어 넣지 않으면 펙틴이 덩어리질 수 있다.
5. 물과 전분을 섞어 넣고 농도가 걸쭉해지도록 끓인다.
6. 블렌더로 믹싱해 냉장고에 보관한다.

Mandarin Compote

280 g	Mandarins
30 g	Mandarin juice
90 g	Sugar
3 g	Pectin
6 g	Water
3 g	Starch

1. Separate the flesh and peel of mandarins.
   - Here, fresh Hallabong fruits were used that are easy to obtain in Korea. Depending on your taste, you can substitute with orange or Cheonhyehyang.
2. Put cold water (other than the requested) and mandarin peels in a pot, and when it boils, strain through a sieve. Repeat this process a total of three times, then finely grind with a blender.
3. Boil mandarin flesh, mandarin juice, and 2 in a pot, then mix with a blender and refrigerate.
   - Be careful not to grind the contents too finely.
4. Stir in previously mixed sugar and pectin, and boil.
   - If the sugar and pectin are not mixed ahead, pectin may become lumpy.
5. Stir in previously mixed water and starch, and boil to thicken.
6. Mix with a blender and store in a refrigerator.

2

3

5

6

금귤 절임

슬라이스한 금귤	10장
만다린즙	20g

슬라이스한 금귤을 만다린즙에 담가 냉장고에 보관한다.

블러드오렌지 소르베

우유	140g
꿀	5g
설탕	60g
글루코스(파우더)	40g
아이스크림 안정제 (stabilizer)	1g
블러드오렌지즙	340g
블러드오렌지 제스트	적당량

1. 냄비에 우유, 꿀을 넣고 데운다.
2. 미리 섞어둔 설탕, 글루코스, 아이스크림 안정제를 넣고 휘퍼로 저어가며 끓인 후 불에서 내려 식힌다.
3. 블러드오렌지즙, 블러드오렌지 제스트를 넣고 블렌더로 믹싱한 후 아이스크림 머신에서 소르베로 만든다.
   ● 아이스크림 머신 대신 파코젯Pacojet을 사용해도 좋다.

Steeped Kumquat

10 slices	Kumquats, sliced
20 g	Mandarin juice

Steep the sliced kumquats in mandarin juice and store in the refrigerator.

Blood Orange Sorbet

140 g	Milk
5 g	Honey
60 g	Sugar
40 g	Glucose powder
1 g	Ice cream stabilizer
340 g	Blood orange juice
QS	Blood orange zest

1. Warm milk and honey in a saucepan.
2. Stir in previously mixed sugar, glucose powder, and ice cream stabilizer with a whisk and boil while stirring, then remove from the heat to cool.
3. Add blood orange juice and blood orange zest, mix with a blender, then make into sorbet with the ice cream maker.
   ● If you don't have an ice cream maker, you can also use Pacojet.

1

2

3

4

5

플레이팅

1. 접시에 사블레 브르통을 올린다.

2. 사블레 브르통 안에 칼라만시 크림을 채운다.

3. 칼라만시 크림 위에 만다린 콩포트를 파이핑한다.

4. 세그먼트한 과일을 올린다.

5. 과일 위에 오렌지 콩피, 레몬 콩피, 금귤 콩피, 금귤 절임, 핑거라임 과육을 올린다.

6. 블러드오렌지 소르베를 올린다.

7. 연꽃 머랭으로 덮어준다.

7

| Plating | 1. | Put a sablé Breton on a plate. |

Plating

1. Put a sablé Breton on a plate.
2. Fill kalamansi cream in the sable.
3. Pipe mandarin compote over the kalamansi cream.
4. Arrange segmented fruits.
5. Put orange confit, lemon confit, kumquat confit, steeped kumquats, and finger lime pulps on top of the fruits.
6. Place blood orange sorbet on top.
7. Cover with lotus meringue.

For me, who loves citrus fruits, winter is a season I always look forward to with excitement. While I was thinking about making desserts using citrus fruits in winter, I thought about making vacherin by combining tangy, sweet, and refreshing fruits with their best match, meringue. Since the dessert called 'Vacherin' has already been made in various ways by many chefs, I thought a lot about how I could create a new one in my own way. In the meantime, I found the idea notes I had written down since my French days, thinking, 'I'll have to try expressing these things as desserts later.' And in it, I found a motif from the idea of lotus lanterns and lotus flowers, and after many attempts, I was able to complete the dessert called Lotus Flower. Since the Star of this dessert is citrus fruits, I tried to use as many different kinds as possible. To do so, I applied various cooking methods using citrus in cream, sorbet, jam, confit, and segments used in this dessert. At the end, it is covered with a thin meringue piped in the shape of a lotus flower and served so that you can enjoy the fun of cracking it with a spoon.

7

/

New York-
Seoul

/

뉴욕-서울

　'베이비 바나나', '블랙 트러플 콘'과 함께 저의 대표 시그니처 메뉴인 '뉴욕-서울'입니다. 뉴욕-서울은 정식당에서도 꾸준한 사랑을 받았던 인기 메뉴인데, 임정식 셰프님을 비롯한 많은 직원들이 가장 좋아하는 메뉴이기도 합니다. 또한 저의 정체성을 가장 잘 보여주는 메뉴라 가장 애착이 가는 디저트이기도 합니다. 대표적인 프렌치 디저트 중 하나인 슈(파리 브레스트), 한국의 현미를 이용한 크림과 캐러멜, 농장에서 온 신선하고 맛있는 피칸, 미국인들이 사랑하는 쿠키 도우, 그리고 프렌치 테크닉을 바탕으로 한국의 식재료와 로컬 재료를 사용해 현지 입맛에 맞춘, 저의 정체성과 너무나도 닮아 있는 디저트라고 생각합니다. 그래서 저에게 더 특별하고 의미가 큰 디저트입니다.

　명칭도 대표적인 프렌치 디저트 중 하나인 '파리 브레스트'에서 모티브를 따 뉴욕-서울이라고 짓게 되었습니다. 어렸을 적 할머니께서는 명절마다 직접 볶은 곡식이나 강정을 싸주셨는데, 그때의 기억을 더듬어 고소한 맛을 극대화하기 위해 현미를 덮어 인퓨징해 무스를 만들었습니다. 쿠키 도우는 제가 좋아하는 미국의 어느 브랜드의 쿠키 도우가 들어간 아이스크림에서 영감을 받아 굽지 않는 옥수수 쿠키 도우를 만들어 그 위에 찐득한 바닐라 아이스크림을 올렸습니다. 이런 저런 테스트를 하던 중, 옥수수가루를 오븐에 두고 잊어버린 적이 있는데 예상 외로 오븐에서 구워진 옥수수가루의 풍미와 향이 너무 좋아 쿠키 도우의 베이스로 사용하게 되었습니다. 유레카였죠! 견과류와 현미의 달콤하면서도 짭조름한 맛, 그리고 고소함을 바탕으로 바닐라와 크림의 부드러움, 피칸과 튀일의 바삭함이 다채롭게 어우러진 식감을 느낄 수 있는, 제가 가장 좋아하는 디저트 중 하나입니다.

Ingredients

브라운 크럼블

버터	40g
황설탕	50g
중력분	44g
코코아파우더	6g

Brown Crumble

40 g	Butter
50 g	Light brown sugar
44 g	All-purpose flour
6 g	Cocoa powder

Directions

1. 볼에 차가운 상태의 깍둑 썬 버터, 황설탕을 넣고 가볍게 믹싱한다.
2. 체 친 중력분, 코코아파우더를 넣고 반죽이 한 덩어리가 될 때까지 믹싱한다.
   - 바삭한 식감을 위해 가루 재료를 넣은 후 너무 오래 믹싱하지 않도록 주의한다.
3. 반죽을 랩으로 싸 냉장고에 둔다.
4. 반죽을 2mm 두께로 밀어 편 후 잠시 냉동실에 둔다.
5. 크고 작은 깍지들을 이용해 자른 후 냉동실에 보관한다.

1. Lightly mix cold butter cut into cubes and light brown sugar in a mixing bowl.
2. Add sifted flour and cocoa powder and mix until the dough comes together into a ball.
   - Be careful not to mix too long after adding the powdered ingredients for a crispy texture.
3. Wrap the dough with plastic wrap and store it in the refrigerator.
4. Roll out the dough to 2 mm thickness and freeze it for a while.
5. Cut the dough with different sizes of piping tips and freeze them.

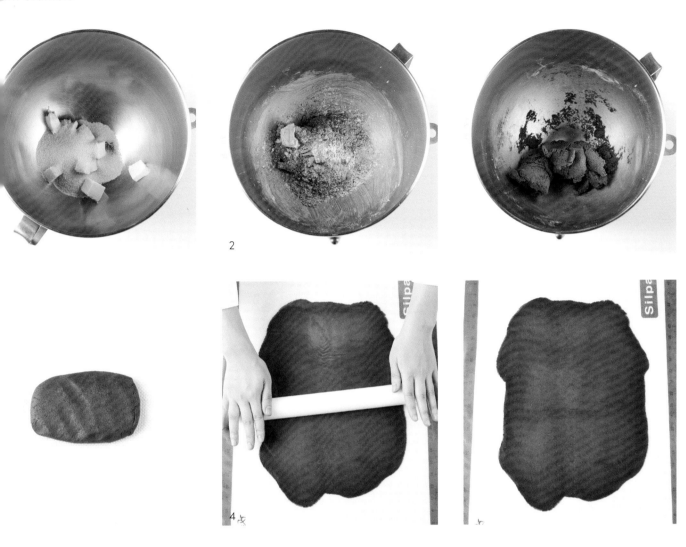

2

4

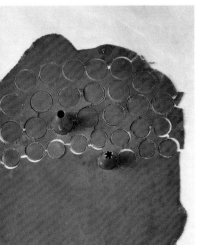

파트 아슈

우유	93g
물	93g
버터	57g
소금	3g
트리몰린	8g
중력분	110g
달걀	3개

1. 냄비에 우유, 물, 버터, 소금, 트리몰린을 넣고 끓어오를 때까지 가열한다.
   - 트리몰린 대신 글루코스나 꿀로 대체할 수 있다.

2. 불에서 내린 후 체 친 중력분을 넣고 섞는다.

3. 다시 불에 올려 반죽이 매끄러워질 때까지 호화시킨다.
   - 반죽이 한 덩어리로 잘 뭉쳐지고 냄비 바닥에 얇은 막이 전체적으로 생기면 불에서 내린다.

4. 믹싱볼에 옮긴 후 달걀을 조금씩 넣어가며 믹싱한다.

5. 원형 깍지(8번)로 실팻 위에 사진과 같이 작은 원형(약 3g), 큰 원형(약 5g)으로 파이핑한다.

6. 브라운 크럼블을 올린다.

7. 170°C로 예열된 오븐에서 25분간 굽는다.

Pâte à Choux

93 g	Milk
93 g	Water
57 g	Butter
3 g	Salt
8 g	Inverted sugar
110 g	All-purpose flour
3	Eggs

1. Heat milk, water, butter, salt, and inverted sugar in a pot until the mixture boils once.
   - You can use glucose or honey instead of inverted sugar.

2. Remove from the heat and mix with sifted all-purpose flour.

3. Put it back over the heat and gelatinize until the dough becomes smooth.
   - Remove from the heat when the dough comes together and a thin film is formed overall on the bottom of the pot.

4. Transfer it to a mixing bowl and beat while adding eggs a little bit at a time.

5. With a round tip (No. 8), pipe small rounds (about 3 grams) and large rounds (about 5 grams) as shown in the photo.

6. Place brown crumble on top.

7. Bake for about 25 minutes in an oven preheated to 170°C.

2

3

5

7

피칸 프랄린

물	50g
설탕	100g
피칸	130g
헤이즐넛	70g
소금	1g
파에테포요틴	40g

1. 냄비에 물, 설탕을 넣고 연한 갈색으로 캐러멜화시킨다.

2. 피칸, 헤이즐넛을 넣고 잘 섞는다.
   - 피칸과 헤이즐넛은 170℃로 예열된 오븐에서 10분 정도 구운 후 사용한다.

3. 피칸, 헤이즐넛, 캐러멜의 색이 충분히 나면 불을 끄고 소금을 뿌린다.

4. 유산지나 실팻 위에 펼쳐 식힌다.

5. 리퀴드한 상태가 될 때까지 믹서로 갈아준다.

6. 파에테포요틴은 가장 마지막에 넣고 간다.
   - 식감을 위해 너무 곱게 갈리지 않도록 주의한다.

7. 짤주머니에 담아 서늘하고 건조한 실온에 둔다.

Pecan Praline

50 g	Water
100g	Sugar
130 g	Pecans
70 g	Hazelnuts
1 g	Salt
40 g	Paillete feuilletine

1. Caramelize water and sugar in a pot until light brown.

2. Add pecans and hazelnuts, and mix well.
   - Use pecans and hazelnuts after baking them for about 10 minutes in an oven preheated to 170°C.

3. When the pecans, hazelnuts, and caramel obtains sufficient color, remove from the heat and sprinkle with salt.

4. Spread them over parchment paper or Silpat to let cool.

5. Ground them in a blender until liquefied.

6. Add paillete feuilletine at the last minute and grind.
   - Be careful not to grind too finely for texture.

7. Put it in a piping bag and store in a cool and dry place.

5

6

볶은 현미 캐러멜

현미	28g
생크림(36%)	100g
우유	35g
글루코스A	50g
게랑드 소금	1g
설탕	40g
글루코스B	20g
버터	35g

1. 팬에 현미를 넣고 타지 않게 냄비를 흔들어가며 갈색 빛이 돌 때까지 볶는다.
2. 생크림, 우유, 글루코스A, 게랑드 소금을 넣고 끓어오를 때까지 가열한다.
3. 다른 냄비에 설탕, 글루코스B를 넣고 가열한다.
4. 185°C까지 캐러멜화시킨 후 **2**를 조금씩 부어가며 106°C까지 가열한다.
5. 체에 거른다.
6. 버터를 넣고 블렌더로 믹싱해 랩을 씌워 냉장고에 보관한다.

Roasted Brown Rice Caramel

28 g	Brown rice
100 g	Heavy cream (36%)
35 g	Milk
50 g	Glucose A
1 g	Guérande salt
40 g	Sugar
20 g	Glucose B
35 g	Butter

1. Roast brown rice in a pot until it turns brown while shaking the pot to prevent it from burning.
2. Add heavy cream, milk, glucose A, and Guérande salt and bring them to a boil.
3. In a separate saucepan, heat sugar and glucose B.
4. Caramelize until 185°C and continue to cook while adding **2** a little bit at a time until it reaches 106°C.
5. Filter through a sieve.
6. Mix butter with a blender, cover with plastic wrap and store in the refrigerator.

sted Brown Rice Caramel

볶은 현미 바닐라 크림

현미	17g
생크림(36%)	200g
바닐라빈	1/2개
소금	한꼬집
화이트초콜릿	43g
VALRHONA IVOIRE(35%)	
젤라틴매스(200bloom)	8g

1. 팬에 현미를 넣고 타지 않게 냄비를 흔들어가며 갈색 빛이 돌 때까지 볶는다.
2. 생크림, 바닐라빈, 소금을 넣고 끓인 후 불에서 내려 10분간 인퓨징한다.
3. 녹인 화이트초콜릿에 **2**를 두 번에 나눠 넣어가며 섞는다.
4. 녹인 젤라틴매스를 넣고 블렌더로 믹싱한다.
5. 체에 걸러 밀착 랩핑해 냉장고에 보관한다.

옥수수 크럼블

버터	52g
설탕	40g
소금	1g
옥수수가루	54g

1. 볼에 차가운 상태의 깍둑썰기한 버터, 설탕, 소금을 넣고 믹싱한다.
2. 체 친 옥수수가루를 넣고 믹싱한다.
3. 믹싱한 반죽은 강판에 갈아 크럼블 덩어리로 만든다.
   - 완성된 크럼블은 철판 위에 펼쳐 냉동실에서 얼린 후 소분해 냉동실에 보관한다.

Roasted Brown Rice Vanilla Cream

17 g	Brown Rice
200 g	Heavy cream (36%)
1/2	Vanilla bean
Pinch of Salt	
43 g	White chocolate
(Valrhona Ivoire 35%)	
8 g	Gelatin mass (200bloom)

1. Roast brown rice in a pot until it turns brown while shaking the pot to prevent it from burning.
2. Add heavy cream, vanilla bean, and salt, and bring the mixture to a boil. Remove from the heat and let it infuse for 10 minutes.
3. Pour **2** into melted white chocolate in two parts and mix.
4. Add melted gelatin and mix with a blender.
5. Filter through a sieve. Cover, making sure the wrap adheres to the cream, and refrigerate.

Corn Crumble

52 g	Butter
40 g	Sugar
1 g	Salt
54 g	Corn powder

1. Mix cold butter cut into cubes, sugar, and salt in a mixing bowl.
2. Mix with sifted corn powder.
3. Grate the mixed dough on a grater to make crumbles.
   - Spread the crumbles on a baking tray to freeze, then divide them into small portions and store in the freezer.

asted Brown Rice Vanilla Cream

2

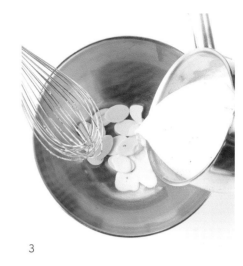

3

4

5

rn Crumble

2

3

피칸 정과

물	15g
설탕	50g
피칸	120g
게랑드 소금	1g
오일	적당량

1. 냄비에 물, 설탕을 넣고 연한 갈색이 될 때까지 캐러멜화시킨다.
2. 피칸을 넣고 진한 갈색이 될 때까지 부서지지 않게 잘 저으며 가열한다.
색이 나면 게랑드 소금을 뿌리고 잘 섞은 후 불에서 내린다.
   ● 피칸은 170℃로 예열된 오븐에서 10분간 구워 사용한다.
3. 유산지나 실팻 위에 오일 스프레이를 살짝 뿌린다.
피칸이 뜨거울 때 장갑을 낀 손으로 하나씩 떼어 실팻 위에서 식힌다.
   ● 오일 스프레이 대신 식용유를 얇게 발 라도 좋다.

피칸 튀일

슈거파우더	38g
박력분	8g
흰자	20g
물	100g
소금	1g
버터	18g
다진 피칸	적당량

1. 볼에 체 친 슈거파우더, 박력분, 흰자를 넣고 골고루 섞는다.
2. 냄비에 물, 소금, 버터를 넣고 버터가 모두 녹을 때까지 가열한다.
3. 1에 2를 조금씩 부어가며 섞는다.
4. 실팻을 깐 철판에 붓고 스패출러로 얇게 펼쳐 170℃로 예열된 오븐에서 5분간 굽는다.
5. 다진 피칸을 골고루 뿌려준 후 20분 정도 더 굽는다.
6. 20분이 지나면 오븐 문을 열고 손바닥 크기로 부신 후 자연스러운 모양으로 구겨 꽃모양을 만든다.
   ● 뜨거울 때 모양을 잡기 쉬우므로 장갑을 끼고 빠르게 작업한다.

Candied Pecans

15 g	Water
50 g	Sugar
120 g	Pecans
1 g	Guérande salt
QS	Oil

1. Caramelize water and sugar in a pot until it turns light brown.
2. Add pecans and heat while stirring without breaking them until it turns dark brown. When the mixture obtains sufficient color, sprinkle with Guérande salt, mix evenly, and remove from the heat.
   ● Use pecans after baking for 10 minutes in an oven preheated to 170℃.
3. Spray a thin layer of oil on parchment paper or a Silpat. Wear gloves, separate each pecans while hot, and let them cool on the Silpat.
   ● You can brush a thin layer of vegetable oil instead of spraying with oil.

Pecan Tuiles

38 g	Powdered sugar
8 g	Cake flour
20 g	Egg whites
100 g	Water
1 g	Salt
18 g	Butter
QS	Pecans, chopped

1. Mix sifted powdered sugar, cake flour, and egg whites evenly in a bowl.
2. Heat water, salt, and butter in a pot until all the butter is melted.
3. Pour 2 into 1 little by little while mixing.
4. Pour over the baking tray lined with a Silpat, spread thinly with a spatula, and bake for 5 minutes in an oven preheated to 170℃.
5. Evenly sprinkle with chopped pecans and bake for about 20 minutes more.
6. After 20 minutes, open the oven door, break it into palm-sized pieces, and crumple them to make flower shapes.
   ● Wear gloves and work quickly, as they are easier to shape while hot.

ndied Pecans

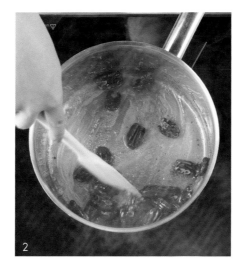

can Tuiles

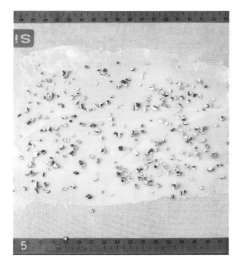

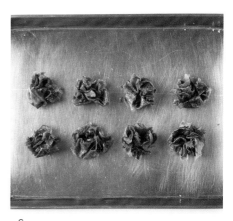

바닐라 아이스크림

우유	278g
생크림(36%)	117g
타히티산 바닐라빈	1/4개
마다가스카르산 바닐라빈	1/4개
설탕	70g
노른자	95g

1. 냄비에 우유, 생크림, 두 가지 바닐라빈을 넣고 끓인 후 불에서 내려 10분간 인퓨징한다.

2. 미리 섞어둔 설탕과 노른자를 넣고 85°C 정도가 될 때까지 휘퍼로 빠르게 저어가며 걸쭉한 상태로 호화시켜 크렘 앙글레이즈를 만든다.

3. 체에 거른다.

4. 블렌더로 믹싱한다.

5. 밀착 랩핑해 하루 동안 냉장 숙성시킨 후 한번 더 블렌더로 믹싱해 아이스크림 기계에서 아이스크림으로 만든다.

   ● 아이스크림 머신 대신 파코젯Pacojet을 사용해도 좋다.

Vanilla Ice Cream

278 g	Milk
117 g	Heavy cream (36%)
1/4	Tahiti vanilla bean
1/4	Madagascar vanilla bean
70 g	Sugar
95 g	Egg yolks

1. Bring milk, heavy cream, and the two kinds of vanilla beans to a boil, then remove from the heat and infuse for 10 minutes.

2. Stir in previously mixed sugar and egg yolks and make crème Anglaise while stirring quickly with a whisk until about 85°C, to a thick, gelatinized state.

3. Filter through a sieve.

4. Mix using a blender.

5. Cover the cream, making sure the wrap adheres to the cream, and refrigerate for a day. Mix with a blender again and make ice cream using the ice cream maker.

   ● If you don't have an ice cream maker, you can also use Pacojet.

2

4

5

1

2

3

4

5

6

플레이팅

1. 접시에 반으로 자른 슈를 올린다.

2. 피칸 정과 조각을 슈 위에 올린 후 가볍게 휘핑한 볶은 현미 바닐라 크림을 슈 위에 파이핑한다.

3. 볶은 현미 바닐라 크림 안으로 피칸 프랄린을 파이핑한다.

4. 슈를 덮어준다.

5. 슈 사이에 옥수수 크럼블을 놓는다.

6. 슈 옆으로 볶은 현미 캐러멜을 파이핑한다.

7. 볶은 현미 캐러멜 옆에 피칸 프랄린을 파이핑한다.

8. 피칸 프랄린과 볶은 현미 캐러멜 사이에 반으로 자른 피칸 정과를 놓는다.

9. 피칸 프랄린 위에 피칸 튀일을, 옥수수 크럼블 위에 바닐라 아이스크림을 올린다.

8   9

Plating

1. Place choux cut in half on a plate.
2. Put pieces of candied pecans on the choux and pipe lightly whipped roasted brown rice vanilla cream on top.
3. Pipe pecan praline inside the roasted brown rice vanilla cream.
4. Cover them with choux.
5. Arrange corn crumbles between the choux.
6. Pipe roasted brown rice caramel next to the choux.
7. Pipe pecan praline next to roasted brown rice caramel.
8. Place the halved candied pecans between the pecan praline and the roasted brown rice caramel.
9. Top with pecan tuiles over the pecan praline and vanilla ice cream over the corn crumbles.

It is my signature menu, along with Baby Banana and Black Truffle Cone. New York-Seoul is a popular menu that has been steadily loved at Jungsik in New York, and it was also a favorite menu for several years for many employees, including Chef Jung-Sik Lim. It is also one of the desserts I am most fond of because it's the menu that best shows my identity. It is based on one of the representative French desserts, choux (Paris-Brest), cream and caramel using Korean brown rice, fresh and delicious pecans from farms, cookie dough Americans love, and French techniques. With Korean heritage, French technique-experience, based in NY, and being a New Yorker; I think this dessert closely resembles my identity as being accommodated to New Yorker's flavor. That's why it is a memorable and meaningful dessert for me as well.

I named it New York-Seoul after the motif of 'Paris-Brest,' one of the representative French desserts. Whenever I went to see my grandmother on holidays when I was young, my grandmother used to pack me grains she pan-roasted herself or gangjeong. I traced the memories of that time and made a mousse by infusing pan-roasted brown rice to maximize the nutty flavor. The cookie dough was inspired by an American brand's cookie dough ice cream I like, and I made no-bake corn cookie dough and topped it with sticky vanilla ice cream. While making various tests, I had left some cornmeal in the oven and forgot about it, but I unexpectedly liked the flavor and aroma of the cornmeal baked in the oven, so I ended up using it as a base for cookie dough. It was Eureka! It is one of my favorite desserts in which you can feel the softness of vanilla and cream and the crispiness of pecans and tuile, based on the sweet and salty taste and savory taste of nuts and brown rice.

8

/

Giwa

/

기와

　'기와'는 제가 프랑스에 있던 시절부터 구상했던 디저트입니다. 한국에 아름다운 문화가 많지만 그중에서도 기와를 볼 때마다 아름답다는 생각이 많이 들었습니다. 사진으로 많이 남겨두기도 했고요. 그래서 나중에 꼭 디저트로 표현해보겠다는 생각을 아이디어 노트에 기록하기도 했습니다. 이 버전은 초창기 1.0 버전이고, 2.0 버전은 2019년 발로나 C3 North America 경기에서 선보이고 준우승을 했던, 모양은 같지만 조금 변형된 다른 버전의 기와입니다. 세계적인 무대에서 한국의 문화와 한국적인 재료를 사용한 디저트를 선보이고 알리고 싶은 마음이 항상 있다 보니 선택하게 된 메뉴이기도 합니다.

　기와 모형은 1차적으로는 초콜릿을 제가 직접 조각해 만들었고, 그 조각한 초콜릿을 바탕으로 식용 실리콘을 이용해 몰드를 직접 제작했습니다. 기와를 표현한 디저트를 만들겠다는 결심을 하고 어떻게 하면 맛까지 완벽하게 표현할 수 있을지를 고민하다가 어릴 적 기억 속 시골 풍경을 떠올렸습니다. 메밀, 불을 지필 때 나는 연기와 냄새, 밭에서 파사삭 밟히는 낙엽과 나뭇가지, 떨어진 꽃잎 등 시골의 풍경과 느낌, 냄새를 상상하며 표현했습니다. 실제로 이 디저트를 드셨던 손님들이 같은 마음으로 공감해주셔서 감사했고 재미있었던 메뉴입니다.

Ingredients

6인분 • SERVES 6

Directions

메밀 초콜릿 크레뮤

생크림(36%)	50g
우유	50g
볶은 메밀	12g
노른자	20g
설탕	10g
다크초콜릿	30g
VALRHONA CARAIBE(66%)	
밀크초콜릿	15g
VALRHONA JIVARA(40%)	

1. 냄비에 생크림, 우유, 볶은 메밀을 넣고 가열한다.
   - 볶은 메밀은 시판 제품을 사용하거나, 생메밀을 팬에서 노릇한 상태가 될 때까지 볶아 사용한다.
2. 끓어오르면 불에서 내려 10분간 인퓨징한다.
3. 볼에 노른자, 설탕을 넣고 섞은 후 **2**를 2~3회 나눠 넣어가며 섞는다.
4. 다시 냄비에 옮긴 후 85°C까지 끓여 크렘 앙글레이즈를 만든다.
5. 체에 거른다.
6. 다크초콜릿, 밀크초콜릿이 담긴 볼에 넣고 섞는다.
7. 블렌더로 유화시킨 후 냉장고에 보관한다.

Buckwheat Chocolate Cremeux

50 g	Heavy cream (36%)
50 g	Milk
12 g	Buckwheat, roasted
20 g	Egg yolks
10 g	Sugar
30 g	Dark chocolate
(Valrhona Caraibe 66%)	
15 g	Milk chocolate
(Valrhona Jivara 40%)	

1. Heat heavy cream, milk, and roasted buckwheat in a pot.
   - You can use commercially available roasted buckwheat or roast buckwheat in a pan until it turns golden brown.
2. Once it boils, remove from the heat and infuse for 10 minutes.
3. Mix egg yolks and sugar in a bowl, add **2** in 2~3 portions to mix.
4. Put the mixture back over the heat and boil until 85°C to make a crème Anglaise.
5. Filter through a sieve.
6. Pour into a bowl with dark chocolate and milk chocolate to mix.
7. Emulsify using a blender and store in the refrigerator.

kwheat Chocolate Cremeux

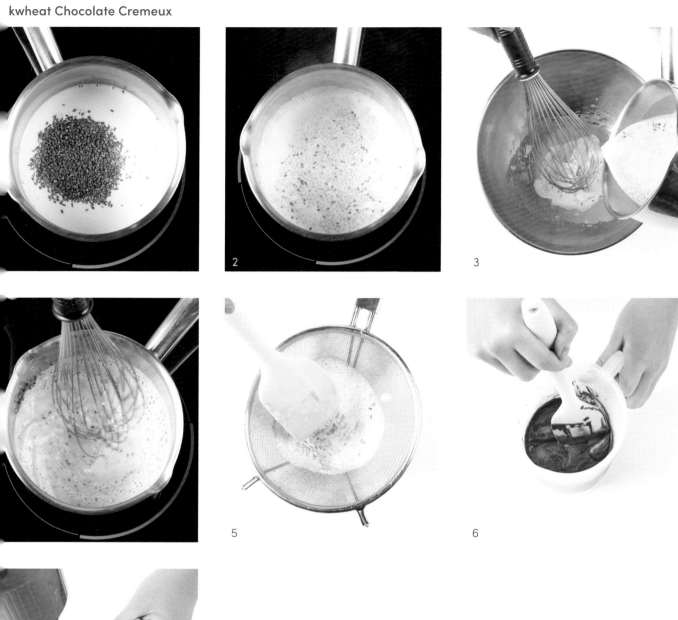

2

3

5

6

훈연 바닐라 아이스크림

우유	240g
황설탕	40g
설탕	18g
글루코스(파우더)	5g
아이스크림 안정제 (stabilizer)	1g
생크림(36%)	100g
바닐라빈	1개

1. 냄비에 우유, 미리 섞어둔 황설탕, 설탕, 글루코스, 아이스크림 안정제를 넣고 끓인다.

2. 볼에 생크림, 바닐라빈을 넣고 10분간 인퓨징한다.

3. 랩을 씌운 후 훈연 기계 호스를 볼 안에 넣고 훈연한다.

   ● 훈연이 끝나면 호스를 빼고 재빨리 랩으로 씌워 15분간 향이 배어들게 한다.

4. 체에 거른다.

5. 1과 함께 블렌더로 유화시킨 후 하루 동안 냉장고에 보관한다. 사용하기 전 한 번 더 블렌더로 믹싱해 아이스크림 머신에서 아이스크림으로 만든다.

   ● 아이스크림 머신 대신 파코젯Pacojet을 사용해도 좋다.

Smoked Vanilla Ice Cream

240 g	Milk
40 g	Light brown sugar
18 g	Sugar
5 g	Glucose powder
1 g	Ice cream stabilizer
100 g	Heavy cream (36%)
1	Vanilla bean

1. Boil milk and previously mixed light brown sugar, sugar, glucose powder, and ice cream stabilizer in a saucepan.

2. Add heavy cream and vanilla bean to a bowl and infuse for 10 minutes.

3. Cover with plastic wrap and insert the smoke machine hose into the bowl and smoke.

   ● After smoking, remove the hose and quickly cover it with plastic wrap, and let the scent soak in for about 15 minutes.

4. Filter through a sieve.

5. Emulsify with 1 with a blender and store in the refrigerator for a day. Mix with a blender once more before use, then make it into ice cream using the ice cream maker.

   ● If you don't have an ice cream maker, you can also use Pacojet.

Smoked Vanilla Ice Cream

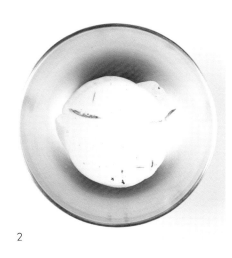

2

3

5

후추 캐러멜

생크림(36%)	37g
우유	23g
후추	약간
게랑드 소금	약간
바닐라빈	1/4개
설탕	53g
글루코스(시럽)	46g
버터	10g

1. 냄비에 생크림, 우유, 후추, 게랑드 소금, 바닐라빈을 넣고 끓인 후 불에서 내려 10분간 인퓨징한다.

2. 다른 냄비에 설탕, 글루코스를 넣고 178℃까지 끓여 캐러멜화시킨다.

3. 체에 거른다.

4. 1과 버터를 넣고 잘 섞는다.

5. 블렌더로 유화시킨 후 냉장고에 보관한다.

기와 초콜릿 1

코코아버터	100g
화이트초콜릿	80g
다크초콜릿	20g
검정색 지용성 색소	0.6g
템퍼링한 다크초콜릿	적당량

1. 볼에 템퍼링한 다크초콜릿을 제외한 모든 재료를 넣고 중탕으로 녹인 후 블렌더로 유화시킨다.

2. 문양이 있는 몰드에 템퍼링한 초콜릿을 파이핑하고 굳힌 후 1을 분사한다.
    - 여기에서는 자체 제작한 몰드를 사용했다.

Pepper Caramel

37 g	Heavy cream (36%)
23 g	Milk
QS	Ground pepper
QS	Guérande salt
1/4	Vanilla bean
53 g	Sugar
46	Glucose (syrup)
10 g	Butter

1. Bring heavy cream, milk, pepper, Guérande salt, and vanilla bean to a boil. Remove from the heat and let infuse for 10 minutes.

2. In a different pot, boil sugar and glucose in a saucepan to 178℃ to caramelize.

3. Filter through a sieve.

4. Combine 1 with butter.

5. Emulsify using a blender and store in the refrigerator.

Giwa Chocolate 1

100g	Cocoa butter
80 g	White chocolate
20 g	Dark chocolate
0.6 g	Black chocolate color
QS	Dark chocolate, tempered

1. Add all the ingredients except tempered dark chocolate in a bowl, melt them over a double boiler, and emulsify with a blender.

2. Pipe tempered chocolate in a mold with a design. When it's crystallized, spray with 1.
    - Here, I used our customized mold.

oper Caramel

4
5

a Chocolate 1

2

기와 초콜릿 2

코코아버터	50g
검정색 지용성 식용 색소	적당량
은색 지용성 식용 색소	적당량
템퍼링한 다크초콜릿	적당량

1. 투명 필름을 4cm 폭으로 자른다. 녹인 코코아버터에 검은색, 은색 지용성 색소를 섞은 후 투명 필름에 얇게 칠한다.
2. 템퍼링한 다크초콜릿을 파이핑한 후 얇게 펴 바른다.
3. 5cm 길이로 자른다.
4. 원통형 실리콘에 넣어 기와 모양이 되도록 굳힌다.
   ● 초콜릿이 충분히 굳으면 투명 필름을 조심스럽게 떼어내어 보관한다.

브라우니

버터	150g
다크초콜릿	80g
VALRHONA GUANAJA(70%)	
설탕	130g
달걀	130g
박력분	35g
코코아파우더	15g

1. 볼에 버터, 다크초콜릿을 넣고 중탕으로 녹인다.
2. 미리 섞어둔 설탕과 달걀을 넣고 섞는다.
3. 체 친 박력분, 코코아파우더를 넣고 섞는다.
4. 반죽을 팬에 붓고(두께 0.7cm) 180°C에서 5~6분간 굽는다.
5. 식은 브라우니는 아래의 사이즈로 자른다.

Giwa Chocolate 2

50 g	Cocoa butter
QS	Black chocolate color
QS	Silver chocolate color
QS	Dark chocolate, tempered

1. Cut the transparent film into 4 cm wide strips. Mix black and silver chocolate coloring with melted cocoa butter and brush thinly on the film.
2. Pipe tempered chocolate over and spread thinly.
3. Cut into 5 cm lengths.
4. Insert them in cylindrical silicone and let them harden in giwa shapes.
   ● Carefully remove the transparent films when the chocolate sets sufficiently.

Brownie

150 g	Butter
80 g	Dark chocolate
(Valrhona Guanaja 70%)	
130 g	Sugar
130 g	Eggs
35 g	Cake flour
15 g	Cocoa powder

1. Add butter and dark chocolate to a bowl and melt them over a double boiler.
2. Combine with previously mixed sugar and eggs.
3. Mix with sifted cake flour and cocoa powder.
4. Pour the batter into a pan (to 0.7 cm thickness) and bake for 5~6 minutes at 180°C.
5. Cut the cooled brownie into the size below.

3

2

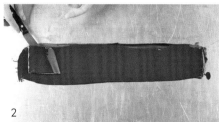

4

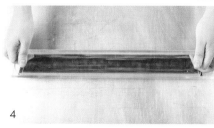

wnie

2

3

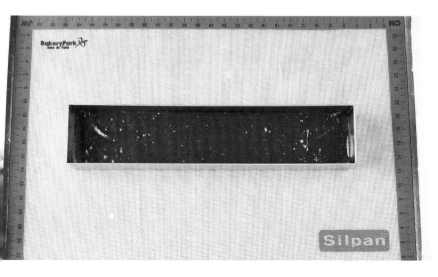

5

메밀 초콜릿 크런치

파에테포요틴	65g
볶은 메밀	40g
소금	1g
다크초콜릿	42g
VALRHONA CARAIBE(66%)	
밀크초콜릿	20g
VALRHONA JIVARA(40%)	
버터	10g

1. 볼에 파에테포요틴, 볶은 메밀, 소금을 넣고 골고루 섞는다.
   - 큰 조각들은 손으로 부순다.
2. 다른 볼에 다크초콜릿, 밀크초콜릿, 버터를 넣고 중탕으로 녹인다.
3. 1과 2를 섞는다.
4. 비닐을 깐 철판 위에 3을 올린 후 그 위에 다시 비닐을 덮고 0.5cm 두께로 밀어 편 후 냉동실에 5분 정도 둔다.
5. 지름 4cm 원형 틀로 자른다.
6. 원형으로 자른 메밀 초콜릿 크런치는 냉동실에 보관한다.
   - 사용하고 남은 메밀 초콜릿 크런치는 살짝 녹인 후 다시 밀어 펴 사용할 수 있다.

메밀 두유 소스

우유	90g
두유(무설탕)	70g
꿀	8g
볶은 메밀	10g

1. 냄비에 모든 재료를 넣고 끓인다.
2. 끓어오르면 중약불로 줄여 반 정도로 졸아들 때까지 가열한다.

Buckwheat Chocolate Crunch

65 g	Paillete feuilletine
40 g	Buckwheat, roasted
1 g	Salt
42 g	Dark chocolate
(Valrhona Caraibe 66%)	
20 g	Milk chocolate
(Valrhona Jivara 40%)	
10 g	Butter

1. Thoroughly combine paillete feuilletine, pan-roasted buckwheat and salt in a bowl.
   - Crumble large pieces with your hand.
2. In a separate bowl, melt dark chocolate, milk chocolate, and butter over a double boiler.
3. Combine 1 and 2.
4. Place the mixture over a transparent film on a baking tray. Cover with another film, roll out to 0.5 cm thickness, and put in a freezer for about 5 minutes.
5. Cut with a 4 cm round cutter.
6. Keep the buckwheat chocolate crunch cut into rounds frozen.
   - Slightly melt the remaining buckwheat chocolate crunch and roll out again to use.

Buckwheat Soy Milk Sauce

90 g	Milk
70 g	Soy milk (sugar-free)
8 g	Honey
10 g	Buckwheat, roasted

1. Boil all ingredients in a pot.
2. Once it boils, reduce the heat to medium-low and simmer until the mixture reduces to about half.

ckwheat Chocolate Crunch

2

3

5

6

ckwheat Soy Milk Sauce

2

메밀 튀일

물	110g
소금	1g
버터	12g
메밀가루	22g
슈거파우더	23g
흰자	27g

1. 냄비에 물, 소금, 버터를 넣고 버터가 녹을 때까지 가열한다.
2. 볼에 체 친 메밀가루와 슈거파우더, 흰자를 넣고 골고루 섞는다.
3. 2에 1을 조금씩 넣어가며 섞는다.
4. 짤주머니에 담고 입구를 좁게 잘라 실팻 위에 둥지 모양으로 파이핑한다.
5. 175℃로 예열된 오븐에서 15분간 굽는다.

Buckwheat Tuile

110 g	Water
1 g	Salt
12 g	Butter
22 g	Buckwheat powder
23 g	Powdered sugar
27 g	Egg whites

1. Heat water, salt, and butter in a pot until the butter melts.
2. In a bowl, combine sifted buckwheat powder and sugar powder, and thoroughly mix with egg whites.
3. Add **1** little by little into **2** to mix.
4. Put the mixture in a piping bag, cut a narrow opening, and pipe in a shape of a nest on a Silpat.
5. Bake for 15 minutes in an oven preheated to 170℃.

3

5

1

2

3

4

5

6

플레이팅

1. 접시 가운데에 브라우니를 세운다.

2. 메밀 초콜릿 크런치 위에 몰딩한 기와 초콜릿 1을 올린 후 브라우니 앞부분에 놓는다.

3. 브라우니 양옆에 메밀 초콜릿 크레뮤를 길게 파이핑한다.

4. 기와 양옆에 후추 캐러멜을 동그랗게 파이핑한다.

5. 기와 옆에 여분의 메밀 초콜릿 크런치를 놓는다.

6. 브라우니 위에 기와 초콜릿 2를 올린다.

7. 메밀 초콜릿 크런치 위에 훈연 바닐라 아이스크림을 올린다.

8. 후추 캐러멜 위에 메밀 튀일을 올린 후 쏘렐 잎을 올린다.

9. 살짝 데운 메밀 두유 소스를 서비스로 준비한다.

8

9

Plating		
	1.	Place a brownie with the slanted side up.
	2.	Put the molded giwa chocolate 1 over buckwheat chocolate crunch and place it in front of the brownie.
	3.	Pipe buckwheat chocolate cremeux along both sides of the brownie.
	4.	Pipe pepper caramel in rounds on both sides of giwa.
	5.	Put extra buckwheat chocolate crunch next to giwa.
	6.	Top the brownie with giwa chocolate 2.
	7.	Put smoked vanilla ice cream on top of the buckwheat chocolate crunch.
	8.	Place buckwheat tuile over the pepper caramel, then top with a sorrel leaf.
	9.	Serve with slightly warmed buckwheat soy milk sauce.

Giwa* is a dessert that I have been planning since I was in France. There are many beautiful cultures in Korea, but especially when I saw giwa, I thought a lot about how beautiful they were. I also took a lot of pictures of them. So, I wrote in the idea note that I'd have to create it as a dessert later, and this was the initial version 1.0. Version 2.0 is different, the same shape but slightly modified, which was the runner-up in the 2019 Valrhona C3 North American Competition. This menu was chosen because I have always wanted to showcase and promote Korean culture and desserts using Korean ingredients on a global stage.

Giwa model was made by sculpting chocolate myself, and I made a mold using a food-grade silicone based on the sculpted chocolate. I decided to make a dessert that expressed giwa, and while thinking about how to compose the taste perfectly, I recalled the countryside from my childhood memories. It was portrayed by imagining the scenery, feeling, and scents of the countryside, such as buckwheat, the smell of smoke from the fire, and the rustling of leaves and branches and fallen petals in the field. In fact, I was grateful that the customers who had this dessert shared empathy from the heart, and it was also a fun menu.

* Giwa is a roofing tile traditionally made with clay, mainly in black with several different designs, usually flowers such as lotus.

9

/

Yuja
Yakgwa

/

유자 약과

'체리 마시멜로(Bloom)'와 함께 밸런타인데이를 마무리하는 특별한 선물로 만들었던 디저트입니다. 우리 조상들이 예로부터 즐겨 먹었던 한국의 전통 과자인 약과는 예쁜 상자에 담아 좋아하는 사람에게 애정을 표현하는 선물로도 사용되었다고 합니다. 그래서 밸런타인데이의 프티 프루Petit fours, 선물용으로 준비해보았습니다. 기름에 튀기는 전통적인 방식의 약과와 다르게 담백하게 구워내는 방식으로 작업했고, 유자를 이용해 단맛을 줄이고 상큼한 맛을 높였습니다.

Ingredients

20개 분량 ◆ 20EA

유자 사블레

박력분	123g
소금	1g
버터	50g
간 생강	1g
꿀	10g
유자청	10g
유자 제스트	적당량
참기름	7g
식용유	5g
다진 레몬 콩피	적당량

Directions

1. 볼에 체 친 박력분, 소금, 차가운 상태의 깍둑 썬 버터를 넣고 모래 알갱이 상태가 될 때까지 믹싱한다.

2. 간 생강, 꿀, 유자청, 유자 제스트, 참기름, 식용유를 넣고 재료가 골고루 섞일 때까지 믹싱한다.

3. 다진 레몬 콩피를 넣고 반죽이 한 덩어리가 될 때까지 믹싱한다.

4. 반죽을 두 덩어리로 나눠 각각 0.4cm 두께로 밀어 편 후 비닐로 싸 냉장고에 보관한다.

5. 반죽 한 덩어리는 지름 3cm 원형 주름 커터로 자른다.

6. 나머지 반죽 한 덩어리를 지름 3cm 원형 주름 커터로 자른 후, 지름 1cm 원형 깍지로 가운데에 한 번 더 구멍을 낸다.

7. 5번 반죽 위에 6번 반죽을 올린 후 냉장고에 보관한다.

8. 180°C로 예열된 오븐에서 15분 정도 굽는다.

Yuja Sablé

123 g	Cake flour
1 g	Salt
50 g	Butter
1 g	Ginger, grated
10 g	Honey
10 g	Yuja preserve
QS	Yuja zest
7 g	Sesame oil
5 g	Vegetable oil
QS	Lemon confit, finely chopped

1. Mix sifted cake flour, salt, and cold butter cut into cubes in a mixing bowl until the mixture resembles coarse sand.

2. Add grated ginger, honey, yuja preserve, yuja zest, sesame oil, and vegetable oil; combine until thoroughly mixed.

3. Add chopped lemon confit and mix until the dough comes together.

4. Divide the dough into two, roll each dough to 0.4 cm thickness, and wrap with plastic to refrigerate.

5. Cut one of the doughs with a 3 cm diameter scalloped cutter.

6. Cut the remaining dough with a 3 cm scalloped cutter, then cut once more in the center with a 1 cm plain round cutter.

7. Place dough **6** on top of dough **5** and refrigerate.

8. Bake in an oven preheated to 180°C for about 15 minutes.

Sablé

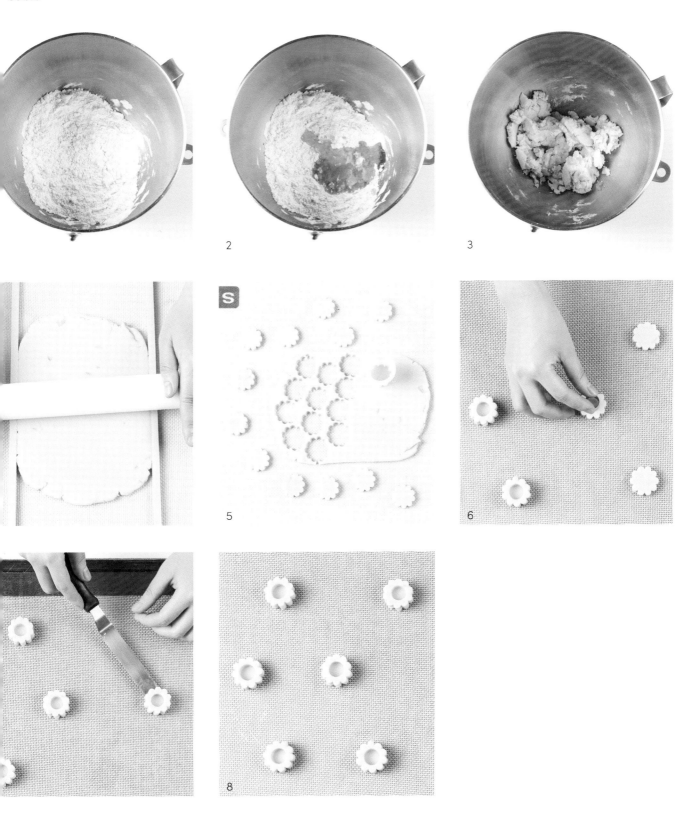

2

3

S

5

6

8

179

유자 시럽

유자청	90g
꿀	65g
메이플시럽	110g
물	50g
유자즙	10g

1. 냄비에 모든 재료를 넣고 끓어오를 때까지 가열한다.
2. 불에서 내려 따뜻한 상태가 될 정도로 식힌 후 구워져 나온 유자 사블레를 5분 정도 담가둔다.
3. 시럽이 사블레에 배어들면 건져내 사용한다.

유자 무침❖

유자 껍질	55g
유자즙	15g
설탕	25g
꿀	5g

1. 볼에 잘게 썬 유자 껍질, 유자즙, 설탕, 꿀을 넣고 골고루 버무린다.
2. 볼 입구를 랩핑해 실온에서 3일간 둔 후 냉장고에 보관한다.

유자 가니시

유자청	10g
유자 무침❖	20g
유자 제스트	적당량

볼에 유자청, 유자 무침, 유자 제스트를 넣고 섞는다.

Yuja Syrup

90 g	Yuja preserve
65 g	Honey
110 g	Maple syrup
50 g	Water
10 g	Yuja juice

1. Heat all ingredients in a saucepan until the mixture boils.
2. Remove from the heat and let cool until it feels warm, and soak the baked sablé for about 5 minutes.
3. When the syrup gets soaked into the sablé, remove it to use.

Yuja Muchim❖

55 g	Yuja peels
15 g	Yuja juice
25 g	Sugar
5 g	Honey

1. Put thinly sliced yuja peels, yuja juice, sugar, and honey in a bowl. Mix with your hands thoroughly.
2. Wrap the bowl and keep it at room temperature for 3 days, then store it in a refrigerator.

Yuja Garnish

10 g	Yuja preserve
20 g	Yuja muchim❖
QS	Yuja zest

Mix yuja preserve, yuja muchim, and yuja zest in a bowl.

Syrup

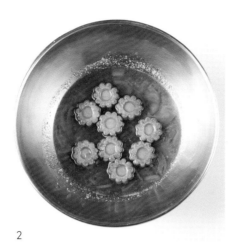

2

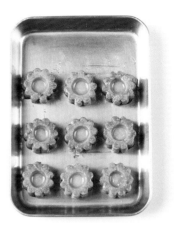

3

Muchim

2

Garnish

1

2

3

플레이팅	1.	접시에 유자 시럽에 담가둔 유자 사블레 2개를 올린다.
	2.	유자 사블레 가운데 빈 공간에 유자 가니시를 채운다.
	3.	식용 금박을 올려 마무리한다.

Plating	1.	Arrange two yuja sables that have been soaked in yuja syrup on a plate.
	2.	Fill the space in the center of the yuja sablé with yuja garnish.
	3.	Decorate with edible gold leaves to finish.

It's a dessert I made as a special present to wrap up Valentine's Day along with cherry marshmallows (Bloom). Yakgwa, a traditional Korean snack our ancestors have enjoyed, is said to have been presented as a gift to express their affection by gifting it to their loved ones in a pretty box. So, I prepared petit fours as a gift for Valentine's Day. Unlike traditional yakgwa, which is fried in oil, this dessert is baked to make it lighter and has reduced sweetness and enhanced fresh taste using *yuja**.

* *Yuja*: Citron, also known as yuzu.

Bloom

봄, 피어남

2020년의 밸런타인데이 콘셉트는 조금 특별했습니다. 밸런타인데이 자체가 서양에서 유래된 문화이긴 하지만 일반적인 하트나 장미 모양의 디저트 말고 뭔가 더 한국적인 느낌을 담을 수 있는 방법이 없을까 고민을 했습니다. 저는 쉬는 날 드라마 보는 것을 좋아하는데요, 그 당시 빠져 있던 사극을 보다가 아이디어를 얻었고, 그래서 그 해의 밸런타인데이의 콘셉트가 '사극'과 '연모하다'가 되었습니다. 제 나름대로 스토리텔링을 해보고 상상하며 메뉴를 구상했습니다. '벚꽃이 만개한 거리를 거닐다 오미자 차를 마시며 이야기를 나누고, 호롱불 아래서 사랑을 나누고, 자갈이 예쁘게 깔린 정원에서 평생 함께 하자는 약속을 하며 약과가 담긴 상자를 선물한다.'는 사극에서나 나올 법한 장면들을 상상하며 메뉴를 만들어보았습니다. 실제로 매장에서는 한국에서 공수한 아름다운 색상의 보자기에 곱게 싸 식사를 마친 커플 손님들을 위한 선물로 준비하기도 했습니다.

페이스트리 셰프로서 밸런타인데이는 꽤 중요한 날입니다. 이 특별한 날마다 매번 일을 해왔지만 항상 손님들 입장에 서서 어떻게 하면 평생 기억에 남는 사랑과 감동이 넘치는 하루를 선사할 수 있을지에 대한 고민을 많이 하게 됩니다. 새벽에 꽃시장에 가 제 키만한 벚꽃 나뭇가지를 낑낑거리며 이고 오기도 했고, 시간 가는 줄 모르고 새벽까지 초콜릿으로 호롱불 스탠드를 만들고, 손글씨와 그림에 탁월한 재능을 가진 팀원들과 함께 메시지 카드에 그림을 직접 그리기도 하고, 정성을 담아 보자기 포장을 하기도 했습니다. 내가 상상한 아이디어가 현실이 될 때의 짜릿함, 손님들에게 어떤 서프라이즈가 될지 기대하는 설렘, 그리고 이 모든 것들을 팀원들과 함께 준비하는 과정과 결과적으로 손님들이 즐거워하는 모습을 보면 정말 뿌듯하고 큰 보람을 느낍니다. 이러한 느낌들이 제가 자부심을 가지며 페이스트리 셰프라는 직업을 계속 할 수 있게 하는 원동력이 되는 것 같습니다. 앞으로도 쭉 손님들에게 감동을 주고 행복을 느끼게 하는 맛있는 디저트를 '계속해서' 만들고 싶습니다.

Ingredients

체리 마시멜로

설탕A	125g
글루코스	20g
체리즙	35g
흰자	45g
설탕B	7g
젤라틴매스	50g
(200bloom)	
오일	적당량

Directions

1. 냄비에 설탕A, 글루코스, 체리즙을 넣고 130°C까지 끓인다.
2. 볼에 흰자, 설탕B를 넣고 맥주 거품 정도로 기포가 올라올 때까지 거품기로 가볍게 저어준다.
3. 2에 1을 조금씩 부어가며 저어준다.
4. 젤라틴매스를 넣고 마시멜로가 36~38°C로 식을 때까지 휘퍼로 빠르게 저어준다.
5. 꽃 모양 몰드에 오일 스프레이를 뿌린다.
   - 여기에서는 플렉시판 미니샬롯(FLEXIPAN ORIGINE FP2071) 실리콘 몰드를 사용했다. 오일 스프레이 대신 식용유를 얇게 발라도 좋다.
6. 마시멜로를 파이핑한다.
7. 파이핑한 마시멜로는 몰드에서 쉽게 떨어질 때까지 약 1시간 정도 냉장고에서 굳힌다.

Cherry Marshmallow

125 g	Sugar A
20 g	Glucose
35 g	Cherry juice
45 g	Egg whites
7 g	Sugar B
50 g	Gelatin mass (200bloom)
QS	Oil

1. Boil sugar A, glucose, and cherry juice in a saucepan until 130°C.
2. In a bowl, lightly whisk egg whites and sugar B until it becomes foamy.
3. Add 1 little by little into 2 while continuing to stir.
4. Pour in melted gelatin mass and whip on high speed until the marshmallow cools down to 36~38°C.
5. Apply oil spray to the flower-shaped mold.
   - Here, I used Flexipan Origine FP2071 silicone mold. You can brush a thin layer of vegetable oil instead of spraying with oil.
6. Fill with marshmallow.
7. Refrigerate to set the piped marshmallows for about one hour until they come off easily from the mold.

2

3

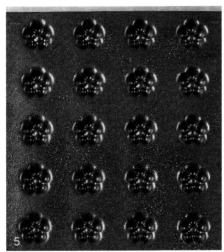

5

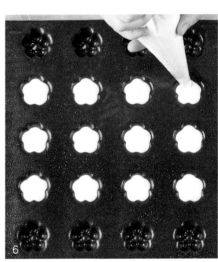

6

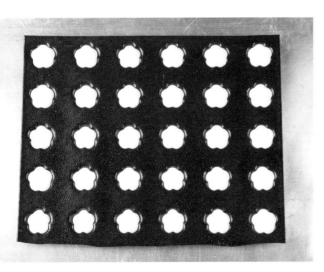

Cherry Sugar

1

2

체리 설탕

설탕	50g
구연산	적당량
동결건조 체리	7g

1. 볼에 설탕, 구연산, 체에 곱게 간 동결건조 체리를 넣고 골고루 섞는다.
2. 마시멜로가 굳으면 몰드에서 분리해 체리 설탕을 묻혀 마무리한다.

플레이팅

준비한 나뭇가지의 뾰족한 부분에 고정시켜 꽃봉오리가 핀 것처럼 연출한다. 나뭇가지 대신 어울리는 접시에 플레이팅해도 좋다.

Cherry Sugar

50 g	Sugar
QS	Citric acid
7 g	Freeze-dried cherries

1. Mix sugar, citric acid, and finely grated freeze-dried cherries thoroughly in a bowl.
2. Once the marshmallows are set, remove them from the mold and coat with cherry sugar to finish.

Plating

Fix onto the tip of the prepared branch to make it look like a flower bud has bloomed. You can plate them on a matching plate instead of using a branch.

The concept for Valentine's Day in 2020 was a little special. Although Valentine's Day itself is a culture that originated in the West, I wondered if there was any way to convey a more Korean touch other than the heart-shaped or rose-shaped desserts that are usually made. I like to watch dramas on my day off, and I was watching a Korean historical-themed drama that I was into at the time, and I slapped my knees and thought, 'This is it!'. So that year's Valentine's Day concept was 'historical K-drama' and 'love.' I did some of my own storytelling and imagination while designing the menu. 'Walking along with the cherry blossoms, sipping omija tea and talking, making love under a lantern, making a promise to be together for life in a garden beautifully decorated with cobblestones, and giving a box of yakgwa as a gift.'; I created a menu imagining scenes that might come out of a historical K-drama. In fact, in the restaurant, yakgwa boxes were wrapped in beautiful colored bojagi (cloth wrapper) imported from Korea and prepared as gifts for couples who finished their meal.

As a pastry chef, Valentine's Day is quite an important day. I have worked every single time on this special day, but I always stand in the customers' shoes and think a lot about how to provide a day full of love and impression to be remembered for a lifetime. At dawn, I went to the flower market and struggled to get branches of cherry blossoms the size of my height, made a lampstand out of chocolate until dawn without realizing the time, drew pictures on message cards with my teammates who have excellent handwriting and drawing skills, and wrapped with bojagi (wrapping cloth) with sincerity. The thrill of turning an idea I imagined into reality, the excitement of anticipating what kind of surprise it will bring to the guests, and the process of preparing all these things together with team members and as a result, I feel proud and rewarded when I see the guests having fun. I think these feelings are the driving force that keeps me proud and able to continue my profession as a pastry chef. I would like to 'continue' making delicious desserts that impress customers and make them happy while feeling this sense of happiness for a long time to come.

LEE EUNJI

To fulfill her dream of becoming a patissier, she went to study in France at the age of eighteen. After completing the Boulangerie course at INBP and the Pasisssrie course at Ecole Ferrandi, she gained various experiences in pastry shops and restaurants in France and New York. She became the first non-European to be the runner-up in season 4 of the popular French TV show <Qui Sera Le Prochain Grand Patissier>, solidifying her position as a pastry chef. Since then, she was selected as a 'Rising Star' by the magazine <Starchefs> with 'Baby Banana,' which she introduced at Jungsik in New York, won the Art of Presentation award, an opportunity to publicize the name 'Eunji Lee' by receiving attention from people around the world on SNS. Currently, she is busy as the owner-chef of the pastry boutique 'LYSÉE' in New York.

@eunji.leeee
@lysee.nyc

2024
Outstanding Pastry Chef and Baker, nominated Semifinalist by James Beard Awards

2023
Pastry Talents of the Year prize by La Liste 1000

Best New Chefs America prize by Food & Wine

Best Original Kouign Amann winner by NYC Kouign-Amann Contest

2022
LYSÉE (New York), Owner-Chef

2019
Valrhona C3 North America Pastry Competition, Second place

<Starchefs> Magazine, Selected as "Rising Star' / Recieved 'Art of Presentation'

2017
French TV show <Qui Sera Le Prochain Grand Patissier>, Season 4, Second Place

2016
Jungsik (New York), Executive Pastry Chef

2012
Hotel Le Meurice (Paris), Chef de Partie

2009
Ze Kitchen Galerie (Paris), In charge of the dessert part

2008
Ecole Ferrandi (Paris), Completed Patisserie course, Certified CAP Patissier

2006
INBP (Rouen) Completed Boulangerie course, Certified CAP Boulanger

PLATING DESSERT

First edition published	August 22, 2022
Third edition published	December 5, 2024

Author	Lee Eunji
Translated by	Kim Eunice
Publisher	Han Joonhee
Published by	iCox, inc.
Plan & Edit	Bak Yunseon
Design	Kim Bora
Photograph	Park Sungyoung
Stylist	Lee Hwayoung
Sales/Marketing	Kim Namkwon, Cho Yonghoon, Moon Seongbin
Management support	Kim Hyoseon, Lee Jungmin

Address	Jomaruro 385beongil 122, Sambo Techno Tower, Unit 2002, Bucheon-si, Kyeonggi-do, Republic of Korea
Website	www.icoxpublish.com
Instagram	@thetable_book
E-mail	thetable_book@naver.com
Phone	82-32-674-5685
Fax	82-32-676-5685
Registration date	July 9, 2015
Registration number	386-251002015000034
ISBN	979-11-6426-218-2 (13590)